Topology Design Methods for Structural Optimization

Topology Design Methods for Structural Optimization

Osvaldo M. Querin
School of Mechanical Engineering, The University of Leeds,
Leeds, United Kingdom

Mariano Victoria
Departamento de Estructuras y Construcción,
Universidad Politécnica de Cartagena, Cartagena, Spain

Cristina Alonso
McKinsey & Company, Madrid, Spain

Rubén Ansola
Departamento de Ingeniería Mecánica,
Universidad del País Vasco, Bilbao, Spain

Pascual Martí
Departamento de Estructuras y Construcción,
Universidad Politécnica de Cartagena, Cartagena, Spain

ELSEVIER

ACADEMIC PRESS
An imprint of Elsevier

Academic Press is an imprint of Elsevier
125 London Wall, London EC2Y 5AS, United Kingdom
525 B Street, Suite 1800, San Diego, CA 92101-4495, United States
50 Hampshire Street, 5th Floor, Cambridge, MA 02139, United States
The Boulevard, Langford Lane, Kidlington, Oxford OX5 1GB, United Kingdom

British Library Cataloguing-in-Publication Data
A catalogue record for this book is available from the British Library

Library of Congress Cataloging-in-Publication Data
A catalog record for this book is available from the Library of Congress

ISBN: 978-0-08-100916-1

For Information on all Academic Press publications
visit our website at https://www.elsevier.com/books-and-journals

Publisher: Mathew Deans
Acquisition Editor: Brian Guerin
Editorial Project Manager: Natasha Welford
Production Project Manager: Mohana Natarajan
Cover Designer: Victoria Pearson

Typeset by MPS Limited, Chennai, India

Dedication

To

George I.N. Rozvany (1930–2015)

Thank you for being a mentor and a friend

Contents

Companion website material for this book is available at https://www.elsevier.com/books-and-journals/book-companion/9780081009161

Preface

Topology optimization continues to establish itself in industry as a very powerful and valuable tool used by engineers for the design and improvement of structures. There is an ever increasing number of commercially available structural analysis software which include topology optimization capabilities as well as exclusive topology optimization software and online tools. There is also an extensive volume of published work which include journal and conference publications and an increasing number of books. So in a way, there was a need for a book which combined the detailed explanations of topology optimization methods with bundled software that would allow the reader to: (1) practice how these methods work and (2) discover what the emerging form of a topology optimized structure is. The aim of this book is therefore to assist the reader to become familiar with topology optimization and to provide a basic set of design tools for the development of two dimensional structures and compliant mechanisms subjected to single and multiple load cases, experiencing linear elastic conditions for single and multimaterial properties. Thus giving the reader first-hand experience of what topology optimization can achieve.

This book is intended for advanced undergraduate and master level students, professors/lecturers, professionals, and researchers in the fields of architecture, engineering and product design. The book will help the reader to understand about the different methods for topology design and how to use them. Professors/lecturers can present the methods in this book to students and show them how topology optimization can be used as a powerful tool to provide innovative structural design solutions. Product designers can use the tools provided without the need to know the intricate detail of the methods, but used them to produce innovative designs of the form of a structure. Researchers can use the book to expand the methods presented to new applications, and professionals can use the tools to find out for themselves about these methods.

The work presented in this book constitutes the collaboration between the authors over the past 13 years from their three research groups, one in the Departamento de Estructuras y Construcción of the Universidad Politécnica de Cartagena in Spain, the second in the Departamento de Ingeniería Mecánica of the Universidad del País Vasco in Spain, and the third in the

School of Mechanical Engineering at The University of Leeds in the United Kingdom.

When we teach topology optimization to students, one of the first question that we ask them is: "*What do holes look like?*" To which we always get the same massed perplexed looked from the students with the obvious answer of "*...well..., a hole looks like a hole...,*" and when prodded a bit further they eventually reveal that "*...all holes are round.*" But when we tell them that a circle is the worst possible shape for a hole, they don't believe us. Only after they have completed the course and have designed a lot of structures using topology optimization, have experienced for themselves how topologies emerge and have started to appreciate that there is no such thing as a round hole in an emerging topology, only then, do they realize that any shape is better than a circle for a hole. By using the software provided in this book, we envisage that the reader will be able to see how the form of a structure will emerge from nothing (using TTO) or be sculptured from a block of material (using SERA.m and liteITD) and reveal what the shape, size and location of all of the cavities and external contour which make up the form of the structure should be. As the famous architect Robert Le Ricolais (1894–1977) stated *"The art of structure is where to put the holes."*

We would like to thank Dr. Pedro Jesús *Martínez Castejón*, for his contribution to the work in the development of the growth method for the size, topology, and geometry optimization of truss structures presented in Chapters 2 and 7, and for his fantastic programming skills in his development of the TTO program provided with this book. We would like to express our deep gratitude to Prof. Dr.-Ing. Dr.-Habil. George I.N. Rozvany (1930–2015) who provided continual support and advice in the development of the SERA method. We would also like to thanks our friends and colleagues and a special thanks to our families for their encouragement and moral support over the past 4 years.

March 2017

Chapter 1

Introduction

1.1 STRUCTURAL OPTIMIZATION (SO)

SO consists of the process of determining the best material distribution within a physical volume domain, to safely transmit or support the applied loading condition(s). To achieving this, the constraints imposed by manufacture and eventual use must also be taken into consideration. Some of these may include increasing stiffness, reducing stress, reducing displacement, altering its natural frequency, increasing the buckling load, manufacture with conventional, or advanced methods.

There are currently three different types of optimization methods which come under the heading of SO, these are: (1) size, (2) shape, and (3) topology optimization[1,2].

In Size optimization [3], the engineer or designer knows what the structure will looks like, but does not know the size of the components which make up that structure. For example, if a cantilever beam was going to be used, its length and position may be known, but not its cross-sectional dimensions (Fig. 1.1A). Another example would be a truss structure where its overall dimensions may be known but not the cross-sectional areas of each truss element (bar), Fig. 1.1B. Yet another example would be the thickness distribution of a shell structure. So basically, any feature of a structure where its size is required, but where all other aspects of the structure are known.

In Shape optimization, the unknown is the form or contour of some part of the boundary of a structural domain [2,4]. The shape or boundary could either be represented by an unknown equation or by a set of points whose locations are unknown (Fig. 1.2).

Topology optimization is the most general form of SO [5]. In discrete cases, such as for a truss structures, it is achieved by allowing the design variables, such as the cross-sectional areas of the truss members, to have a value of zero or a minimum gage size (Fig. 1.3). For continuum-type structures in two dimensions (2D), topology changes can be achieved by allowing the thickness of a sheet to have values of zero at different locations, thereby determining the number and shape of the cavities (holes). For continuum-type structures in three-dimensions (3D), the same effect can be achieved by having a density-like variable that can take any value down to zero.

Topology Design Methods for Structural Optimization.
DOI: http://dx.doi.org/10.1016/B978-0-08-100916-1.00001-5

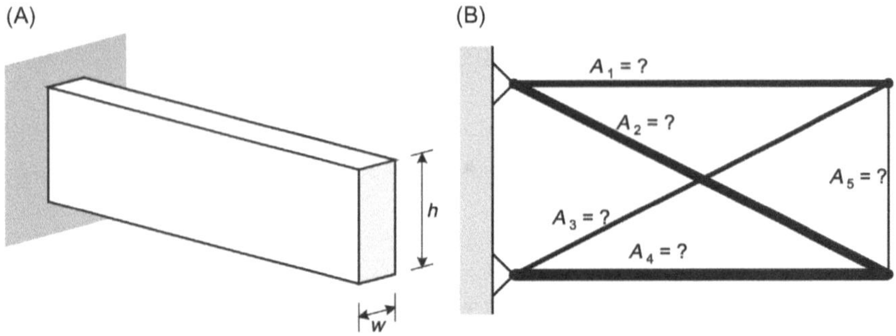

FIGURE 1.1 Examples of structures where size optimization can be used: (A) cantilever beam of unknown cross-sectional dimensions, (B) truss structure where the area of each bar is unknown.

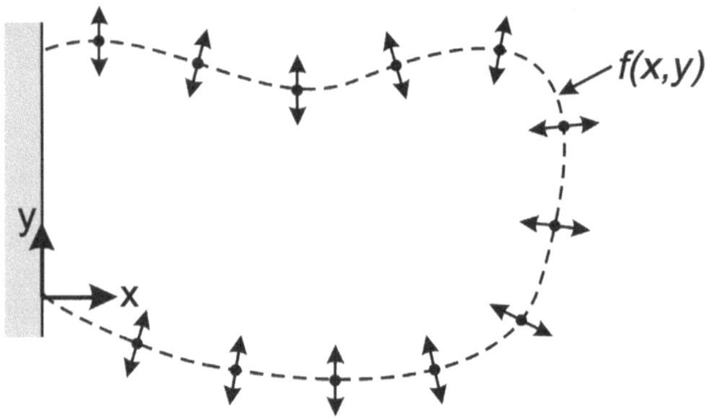

FIGURE 1.2 Structural design domain with the boundary represented as either an equation $f(x,y)$ or as control points which can move perpendicularly (or otherwise) to the boundary.

Alternatively, elements of a structure, such as the finite elements (FE) used to represent it, can be removed or added to the domain (Fig. 1.4).

1.2 TOPOLOGY OPTIMIZATION

In size and shape optimization, the size and shape of the components of a structure can be manipulated. They can have any value between their limits, but they must always be present. But if the designer/engineer does not know what the shape or size of the structure should be, then topology optimization needs to be used. The two major distinctive features of topology optimization are that: (1) the elastic property of the material, as a function of its density, can vary over the entire design domain; and (2) material can be permanently removed from the design domain. There are several topology optimization methods which can be grouped into two categories: (1) Optimality Criteria methods [6,7] and (2) Heuristic or Intuitive methods.

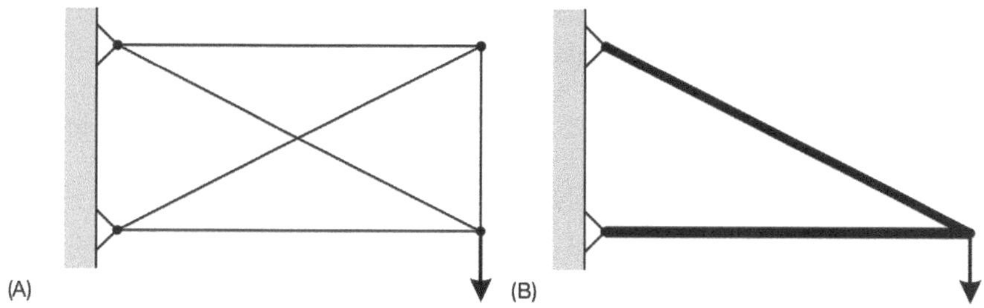

FIGURE 1.3 Topology optimization of a truss structure: (A) original topology; (B) final topology with some trusses removed.

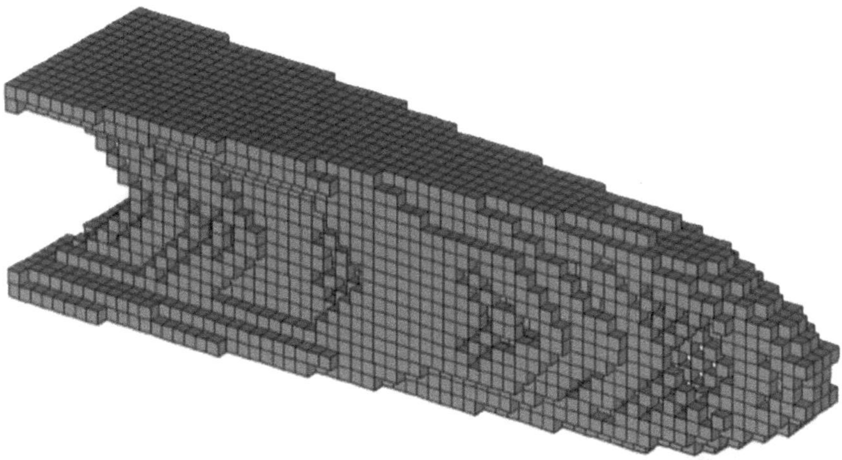

FIGURE 1.4 Final topology for a cantilever beam.

Optimality Criteria are indirect methods of optimization. They satisfy a set of criteria related to the behavior of the structure. They are often based on the Kuhn–Tucker optimality condition [3], which means that they are more rigorous. They are suitable for problems with a large number of design variables and a few constraints. The Optimality Criteria topology methods are: (a) Homogenization [8–10]; (b) Solid Isotropic Material with Penalization (SIMP) [1,8,11]; (c) Level Set Method [12–14]; and (d) Growth Method for Truss Structures (Chapter 2, Growth Method for the Size, Topology, and Geometry Optimization of Truss Structures).

Heuristic methods are derived from intuition, observations of engineering processes, or from observation of biological systems. These methods cannot always guarantee optimality, but can provide viable efficient solutions. Some Heuristic topology optimization methods are: (a) Fully Stressed Design [3] (b) Computer-Aided Optimization (CAO) [15,16]; (c) Soft Kill Option;

(d) Evolutionary Structural Optimization (ESO) [17,18]; (e) Bidirectional ESO (BESO), [19,20]; (f) Sequential Element Rejection and Admission (SERA) (Chapter 3, Discrete Method of Structural Optimization); (g) Isolines/Isosurfaces Topology Design (ITD) (Chapter 4, Continuous Method of Structural Optimization).

A brief review of some of the above mentioned topology optimization methods is given in the following sections.

1.2.1 Homogenization Method for Topology Optimization

The Homogenization method of topology optimization consists of solving a class of shape optimization problem where the topology is made from an infinite number of microscale voids which produces a porous structure [8–10]. The optimization problem then consists of finding the optimum values for the geometry parameters of the microvoids, which become the design variables. If a portion of the structure consists only of voids, material is not placed in that area. Alternatively, this can be thought of as a cavity emerging in that area. This is the reason of why this is classified as a topology optimization method. If a portion of the structure has no porosity, then this corresponds to solid material. Two questions then need to be answered: (1) What do these microvoids look like? and (2) How do they populate the structural domain?

Topology optimization methods are applied to a continuous design domains. These structures have to be evaluated to determine their effectiveness, and an effective way of achieving this is by using FE. If a regular fixed grid FE domain is used with square cells in 2D, then each of these square cells can have one type of microstructure for isotropic materials. The unit cell for an isotropic material structure is represented with a rectangular void of width (a), height (b), and orientation (θ) (Fig. 1.5). The optimization problem is to find the structure with the maximum global stiffness or minimum mean compliance, which is equivalent to finding the maximum potential energy. The optimization problem is then given by Eq. (1.1) [2] where $\overline{V}_s$ is the upper volume limit; N is the number of FE; $\Pi(u)$ is the potential energy in the structure; a_e is the width and b_e is the height of the void, θ_e is the orientation and v_e is the volume the eth FE.

$$\begin{aligned}
&\text{Maximize}: && \Pi(u) \\
&\text{Subject to}: && \sum_{e=1}^{N}(1-a_e b_e)v_e - \overline{V}_s \leq 0 \\
&\text{and} && a_e - 1 \leq 0 \\
& && -a_e \leq 0 \\
& && b_e - 1 \leq 0 \\
& && -b_e \leq 0 \\
& && a_e, b_e, \theta_e : e = 1, 2, \ldots, N
\end{aligned} \tag{1.1}$$

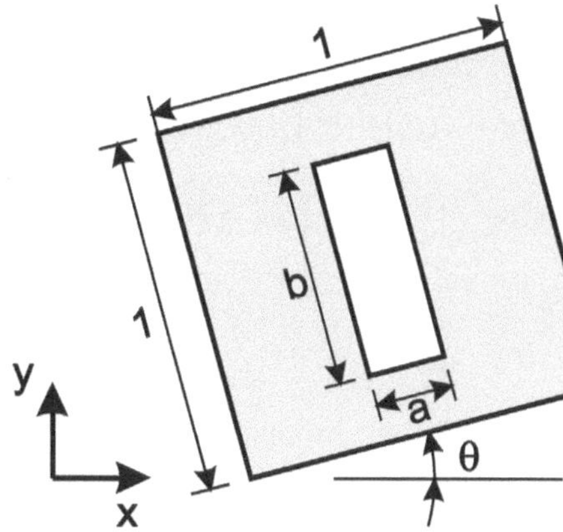

FIGURE 1.5 Unit cell with material and void domain of size $a \times b$ at orientation θ.

1.2.2 Solid Isotropic Material with Penalization (SIMP)

A direct consequence of the Homogenization method, was the development of the SIMP method [1,8,11]. It has since become the most published and implemented in commercial software method of topology optimization. The idea is to use only one design variable per FE. This design variable is an artificial element density (ρ_e) with any value in the range of $0 < \rho_{\min} \leq \rho_e \leq 1$. This means that the volume of an element is multiplied by this artificial density to produce its actual volume, so that the volume of the design domain is given by (1.2).

$$V = \sum_{e=1}^{N} v_e \rho_e \tag{1.2}$$

where N is the total number of FE used to represent the design domain; ρ_e is the artificial density of the eth element where $0 < \rho_{\min} \leq \rho_e \leq 1$; V is the resulting volume of the design domain.

A penalty (p) is placed on the density when it is multiplied to the elastic modulus of the element Eq. (1.3). As the penalization is increased from one to higher values, the optimal solution produces a discrete valued (0−1) problem.

$$E_e = \rho_e^p E_e^0 \tag{1.3}$$

where E_e^0 is the original material elastic modulus and E_e is the new artificial elastic modulus of the eth element; p is the penalty factor which converts the "*grey*" problem to a black-and-white (0−1) design. In order to obtain true (0−1) designs, values of $p > 3$ are usually required.

The SIMP optimization problem is then given by Eq. (1.4), with the optimization process running iteratively. Firstly, the optimum is calculated for

$p = 1$, then this is increased and the optimization repeated until a (0−1) design is generated.

$$\begin{aligned}
&\text{Maximize:} && c(\rho_e) = \{F\}^T(u) \\
&\text{Subject to:} && \left[\sum_{e=1}^{N} \rho_e^p K_e\right]\{u\} = \{F\} \\
& && \sum_{e=1}^{N} v_e \rho_e \leq \overline{V}_s \\
& && 0 < \rho_{\min} \leq \rho_e \leq 1 : e = 1, 2, \cdots, N \\
& && p = 1, 2, \cdots, p_{\max} : p_{\max} > 3
\end{aligned} \tag{1.4}$$

1.2.3 Fully Stressed Design (FSD)

The FSD method is a very intuitive method of size and topology optimization, which is applicable to structures subject to stress and minimum gage constraints. The FSD optimality criterion states that: "*For an optimum design, each member of a structure that is not at its minimum gage must be fully stressed under at least one of the design load conditions*," [3]. This optimality criterion implies that material should be removed from elements of the structure which are not fully stressed, unless limited by minimum gage constraints. But it requires the explicit assumption that by adding or removing such material it only changes their stresses, with negligible or no effect on the rest of the structure. The following four steps describe how the FSD method works:

1. A structure is divided into ***N*** elements (trusses, beams, or plate FE), giving a total of ***N*** design variables.
2. An allowable stress limit (σ_{Allow}) is specified which could either be the same or different for tension or compression elements of the structure.
3. The stresses in the structure are calculated at the current iteration (i) number, by any means (analytical analysis, FEA, etc.)
4. The value of each design variable for the next iteration ($i + 1$) is calculated using the updating scheme of Eq. (1.5).

$$x_j^{i+1} = x_j^i \times F_j^i \tag{1.5}$$

$$F_j^i = \max\left(\frac{\sigma_j^i}{\sigma_j^{\text{allow}}}\right) \tag{1.6}$$

where i is the current ith iteration number; j is the design variable number ($j = 1, \cdots, N$); x_j^i is the current value of the jth design variable; σ_j^i is the current stress level for the jth design variable; σ_j^{Allow} is the allowable stress limit for the jth design variable; F_j^i is the value of the ratio between the current and allowable stresses for the jth design variable Eq. (1.6). If there is more than one stress value for a design variable, the one that returns the highest value of F_j^i is used; x_j^{i+1} is the new value of the jth design variable for the next iteration $(i+1)$. If the new design variable value (x_j^{i+1}) is smaller than the minimum specified gage value $(x_{\min})$, then the minimum is used, given by: if $x_j^{i+1} < x_{\min}$ then $x_j^{i+1} = x_{\min}$. An equivalent maximum gage limit can also be imposed on all design variables $(x_{\max})$.

5. If the structure is statically determinate, the updating scheme of Eq. (1.5) will give an exact solution in one iteration. Alternatively, steps 3 and 4 are repeat until the stresses have converged to a desired tolerance.

1.2.4 Computer-Aided Shape Optimization (CAO)

Mattheck [15] developed the method CAO which simulates biological growth by volumetrically swelling a structure according to its stress distribution. The swelling process simulates adding material to the structure and the shrinkage process, or negative swelling, simulates the process of removing material from the structure. The swelling can easily be achieved using FEA by using a pseudothermal stress distribution [15,16]. The method works iteratively by analyzing and modifying the structure. The process works in the following way:

1. Define two structural FE models:
   - **a.** The original FE model representing the structural design domain with appropriate supports and loading;
   - **b.** A model for the thermal swelling process with a separate set of supporting conditions that will allow for the desired swelling, while constraining growth to within the structural boundaries. Its elastic modulus is set at 1/400 of its initial value in the growth layer. The previous mechanical load is set at zero. Only the soft surface layer of the component is given a thermal expansion factor greater than zero. The result is a soft outer layer which is hot at the over-loaded regions and relatively cold in the lightly loaded regions.
2. The stress distribution of model (a) is calculated.
3. The stress distribution from step (2) is used to calculate the artificial temperature distribution using Eq. (1.7).

$$T_n = T_{\text{ref}} + k_1 \frac{(\sigma_n - \sigma_{\text{ref}})}{\sigma_{\text{ref}}} \tag{1.7}$$

where n is the nth node of the FE mesh; T_{ref} is the reference temperature of the structure about which swelling and shrinking takes place; σ_n is the stress for the nth node in the FE mesh of the structure; σ_{ref} is reference target stress for the structure; k_1 is the scaling factor, a typical value is 100; T_n is the applied artificial temperature to the nth node of the FE mesh in the growth layer.

4. The pseudothermal loading from step (3) is applied to model (b) and a thermal FEA is carried out.
5. The nodal coordinates of both FEA models (a and b) are updated using the displacements induced by the artificial swelling of step (4). To ensure that these displacements result in only an appropriately small change in the topology, they are scaled by an appropriate amount using Eq. (1.8).

$$[x_n, y_n, z_n]^T_{i+1} = [x_n, y_n, z_n]^T_i + k_2[u_n, v_n, w_n]^T_i \quad (1.8)$$

where i is the ith iteration; T means the transpose of the nodal coordinate or displacement vector; $[x_n, y_n, z_n]^T_i$ are the nodal coordinates and $[u_n, v_n, w_n]^T_i$ are the nodal displacements for the nth node in the ith iteration; k_2 is an appropriate scaling factor determined for the specific problem; $[x_n, y_n, z_n]^T_{i+1}$ are the new coordinates for the nth node

6. Steps 2 to 5 are repeated until a uniform stress state is obtained on the surface of the structure.

1.2.5 Soft Kill Option (SKO)

The SKO method of topology optimization simulates the process of adaptive bone mineralization by varying the elastic modulus of a loaded structure according to its stress distribution [21]. The process works in the following way:

1. Define a structural domain to be optimized and apply boundary conditions and loads. To start, the entire structure is given the same target maximum elastic modulus $E_{\max}$.
2. The stress distribution of the structure is calculated.
3. The stress distribution from step (2) is used together with Eq. (1.9) to modify the elastic modulus of every element of the structure.

$$E^e_{i+1} = E^e_i + k(\sigma^e_i - \sigma_{\text{ref}}) \quad (1.9)$$

where i is the current ith iteration number; E^e_i is the elastic modulus and σ^e_i is the average stress for the eth FE in the current ith iteration; σ_{ref} is the global reference stress which controls the variation of the elastic modulus; k is the scaling factor, a typical value is 1000; E^e_{i+1} is the elastic modulus of the eth FE in the next iteration. This value is limited to the following range $E_{\min} \le E^e_{i+1} \le E_{\max}$ where $E_{\min} = E_{\max}/1000$; $E_{\max}$ is the maximum allowable elastic modulus and is equal to the desired value given to the structure in step (1).

4. Repeat steps 2 and 3 until there is a clear distinction between the regions of high and low elastic modulus.

1.2.6 Evolutionary Structural Optimization (ESO)

The ESO method follows the concept that the slow removal of inefficient material from a design domain will eventually lead to an optimum design. The method is very simple and follows both what a designer does (eliminate wasteful material) and also the principle of a FSD. The ESO method works in the following way:

1. A maximum design domain is defined for the structure and is divided with an FE mesh and all boundary constraints, loads, and material properties are applied.
2. The criteria used to optimize the structure are applied: von Mises stress, displacement, frequency or buckling sensitivity, thermal, etc.
3. Analyze the structure using FEA or other method.
4. All elements in the structure which satisfies the ESO inequality of Eq. (1.10) are removed from the structural domain.

$$\sigma_e \leq RR_s \times \sigma_{\max} \tag{1.10}$$

where s is the sth steady state number; σ_e is the stress in the eth FE of the structure; $\sigma_{\max}$ is the maximum stress in the current structure; RR_s is the rejection ratio in the sth steady state, calculated using Eq. (1.11). A typical initial value is $RR_0 = 0.01$.

$$RR_{s+1} = RR_s + ER \tag{1.11}$$

where ER is the evolutionary rate, a typical value is 0.01.

5. Steps 3 and 4 are repeated until no more elements satisfy Eq. (1.10), then go to step 6.
6. Since no more elements can be removed, a steady state is reached. The rejection rate is updated using Eq. (1.11) and $s = s + 1$.
7. Steps 3 to 6 are repeated until:
   a. The maximum rejection ratio is reached, which typically is 0.25; or
   b. A predetermined volume fraction is reached, which typically is 0.3 (30%).

1.2.7 Bidirectional ESO (BESO)

One problem with the ESO method is that it is one-directional. Once material has been removed, it cannot be reintroduced into the domain. A means of addressing this problem was to allow material to be introduced, or reintroduced into the design domain. The regions where material is to be introduced is around elements which are highly stressed. The process is then very similar to ESO, apart from the addition process. The BESO method works in the following way:

1. A maximum design domain is defined for the structure and is divided with an FE mesh and all boundary constraints, loads and material properties are applied.
2. The BESO method allows the user to specify a minimum number of FE which connect the loads to the supports, or to fill the entire initial design domain with FE. All FE not included in the initial starting domain need to be present but deactivated so that they can be reintroduced into the domain at a later stage.
3. The criteria used to optimize the structure are applied: von Mises stress, displacement, frequency or buckling sensitivity, thermal, etc.
4. Analyze the structure using FEA or other method.
5. All elements in the structure which satisfy the ESO inequality of Eq. (1.10) are removed from the structural domain.
6. All elements in the structure which satisfy the BESO inequality Eq. (1.12) have material added around them.

$$\sigma_e \geq IR_s \times \sigma_{\max} \tag{1.12}$$

where IR_s is the inclusion ratio in the sth steady state, calculated using Eq. (1.13). A typical initial value is $IR_0 = 0.99$.

$$IR_{s+1} = IR_s - EIR \tag{1.13}$$

where EIR is the evolutionary inclusion rate. A typical value is 0.01.
7. Steps 4 to 6 are repeated until no more elements satisfy Eqs. (1.10) and (1.12), then go to step 8.
8. Since no more elements can either be added or removed from the design domain, a steady state is reached. Update the rejection and inclusion rates using Eqs. (1.11) and (1.13) and $s = s + 1$.
9. Steps 4 to 8 are repeated until a predetermined volume fraction of the maximum design domain is reached.

Since the initial development of the BESO method there have been several modifications to the way that it works, these can be found in [19].

1.3 BOOK LAYOUT

This book introduces the reader to different methods of topology optimization. The methods explained range from those developed by the authors of this book to other methods developed by other research teams. This book consists of seven chapters organized into four sections: (a) Introduction of structural and topology optimization (this chapter); (2) Description of topology optimization methods applicable to structures represented by: (i) truss elements (Chapter 2, Growth Method for the Size, Topology, and Geometry Optimization of Truss Structures), (ii) discrete FE (Chapter 3, Discrete Method of Structural Optimization) and (iii) continuous representation of a structure (Chapter 4, Continuous Method of Structural Optimization) which were developed by the authors of this book; (3) Explanation of how topology optimization is used to generate structures and what features emerge as a consequent of topology optimization (Chapters 5 and 6, Hands-On Applications of Structural Optimization and Topology Optimization as a Digital Design Tool); and (4) Description of the software included with this book together with several examples to illustrate their use (Chapter 7, User Guides for Enclosed Software).

A general introduction into SO, leading to an explanation of topology optimization is given in this chapter. This chapter also provides brief descriptions of other well-known topology optimization methods such as: Homogenization; SIMP, FSD; CAO; SKO; ESO; and BESO.

Chapter 2, Growth Method for the Size, Topology, and Geometry Optimization of Truss Structures, presents the growth method for the optimal design of statically determinate plane truss structures and also provides a brief introduction to the optimization of truss structures. The algorithm which describes how joints and bars are introduced into the design domain is explained, together with the different optimization steps required for the method to work: topology, size, and geometry.

Unlike other discrete methods of topology optimization, the SERA method uses two separate criteria, one for the "*real*" material and the other for the "*virtual*" material. Chapter 3, Discrete Method of Structural Optimization, describes this method, how it came to be developed, and its applicability to structural and compliant problems.

The ITD method is presented in Chapter 4, Continuous Method of Structural Optimization, where an explanation of its inner workings is given. The extensions of the method for problems with: different tensile and compressive structural behavior, different load case, and different material phases is also given.

When someone starts to learn about topology optimization, there are "*classical*" problems which they should always optimize. This is not only to become familiar with the software used, but also to learn to identify the different characteristics that emerge within an optimized structure. Chapter 5, Hands-On Applications of Structural Optimization, provides the reader with this set of "*classical*" problems.

Topology optimization can be used to modify existing designs, to incorporate explicit features into a design and to generate completely new designs. However, this has mostly only been appreciated by structural designers and engineers. Chapter 6, Topology Optimization as a Digital Design Tool, helps the reader understand how to use topology optimization as a digital tool by examining nine different examples.

Three different programs were developed to demonstrate the methods presented in Chapters 2–4. These programs are: (1) TTO, which implements the Growth Method for truss structures; (2) "*SERA.m*" which is a MATLAB program of the SERA method which allows the user to manipulate the program and experiment with the SERA method; and (3) liteITD which implements the ITD method. Chapter 7, User Guides for Enclosed Software, provides a different user guide for each of the three programs and shows the user how the programs can be used to optimize different topologies.

REFERENCES

[1] M.P. Bendsøe, O. Sigmund, Topology optimisation, Theory, Methods and Applications, Springer-Verlag, Berlin, 2003.

[2] B. Hassani, E. Hinton, Homogenization and Structural Topology Optimization. Theory, Practice and Software, Springer, London, 1999, ISBN: 3-540-76211-6.

[3] R.T. Haftka, Z. Gürdal, Elements of Structural Optimization, third ed., Kluwer, Dordrecht, 1992.

[4] J. Sokolowski, J.P. Zolesio, Introduction to shape optimization: shape sensitivity Analysis, Springer Ser. Comput. Math., vol. 10, Springer, Berlin, 1992.

[5] G.N. Rozvany, M.P. Bendsøe, U.U. Kirsch, Layout Optimization of Structures, ASME. Appl. Mech. Rev. 48 (2) (1995) 41–119. Available from: http://dx.doi.org/10.1115/1.3005097.

[6] M. Zhou, G.I.N. Rozvany, DCOC: an optimality criterion method for large systems. Part I: theory, Struct. Optim. 5 (1992) 12–25.

[7] M. Zhou, G.I.N. Rozvany, DCOC: an optimality criterion method for large systems. Part II: algorithm, Struct. Optim. 6 (1993) 250–262.

[8] M.P. Bendsøe, Optimal shape design as a material distribution problem, Struct. Optim. 1 (4) (1989) 193–202. Available from: http://dx.doi.org/10.1007/BF01650949.

[9] M.P. Bendsøe, Optimization of Structural Topology, Shape and Material, Springer-Verlag, Berlin, 1995.

[10] M.P. Bendsøe, N. Kikuchi, Generating optimal topologies in structural design using a homogenization method, Comput. Math. Appl. Mech. Eng. 71 (1988) 197–224.

[11] G.I.N. Rozvany, M. Zhou, Applications of the COC algorithm in layout optimization, in: H.A. Eschenauer, C. Mattheck, N. Olhoff (Eds.), Engineering Optimization in Design Processes. Lecture Notes in Engineering, vol. 63, Springer, Berlin, Heidelberg, 1991.

[12] G. Allaire, F. Jouve, A.-M. Toader, Structural optimization using sensitivity analysis and a level-set method, J. Comput. Phys. 194 (1) (2004) 363–393.

[13] S. Amstutz, H. Andrä, A new algorithm for topology optimization using a level-set method, J. Comput. Phys. 216 (2) (2006) 573–588.

[14] M.Y. Wang, X. Wang, D. Guo, A level set method for structural topology optimization, Comput. Methods Appl. Mech. Eng. 192 (1–2) (2003) 227–246.

[15] C. Mattheck, Engineering components grow like trees, Mat.-wiss. Werkstofftech 21 (1990) 143–168. Available from: http://dx.doi.org/10.1002/mawe.19900210403.

[16] C. Mattheck, M. Scherrer, K. Bethge, I. Tesari, Shape optimization: an analytical approach, Computer Aided Optimum Design in Engineering IX, WIT Press, Southampton, 2005.

[17] Y.M. Xie, G.P. Steven, A simple evolutionary procedure for structural optimization, Comput. Struct. 49 (1993) 885–896.

[18] Y.M. Xie, G.P. Steven, Evolutionary Structural Optimization, Springer-Verlag, Berlin, 1997.

[19] X. Huang, M. Xie, Evolutionary Topology Optimization of Continuum Structures: Methods and Applications, John Wiley & Sons, Ltd, Sussex, UK, 2010.

[20] O.M. Querin, G.P. Steven, Y.M. Xie, Evolutionary structural optimization using a bidirectional algorithm, Eng. Comput. 15 (1998) 1031–1048.

[21] A. Baumgartner, L. Harzheim, C. Mattheck, SKO (soft kill option): the biological way to find an optimum structure topology, Int. J. Fatigue 14 (6) (1992) 387–393.

Chapter 2

Growth Method for the Size, Topology, and Geometry Optimization of Truss Structures*

2.1 INTRODUCTION

Topology optimization of trusses is a classical subject in structural design. The study of the fundamental properties of optimal grid-like continua was made by Michel [1], and modern layout theory was founded by Prager and Rozvany [2,3].

The most commonly used method for the topology design of truss structures is the Ground Structure approach [4]. This method consists of generating a fixed grid of joints and adding bars in some or all of the possible connections between the joints with each bar acting as a potential structural or vanishing member. The optimum structure for the imposed boundary conditions and applied loads is found using the cross-sectional areas as design variables, including the possibility of removing bars by having a lower limit area of zero. The number of joints is not used as a design variable, so this approach can be considered as a standard sizing problem, although it has the effect of producing a topology.

The problem of size optimization is commonly formulated as a linear programming problem with stress constraints, where the compatibility conditions are suppressed [4–7]. The Simplex formulation of these problems can be seen in Refs. [8,9]. Formulations for displacements only and other formulations in terms of stress only were considered in Refs. [10–13]. Stress and displacement constraints were considered in Refs. [14–16] using the DCOC method. In the classical formulation of this type of problem, the positions of the joints are not used as design variables, so a high number of joints are used to obtain the optimal topology. When a low number of joints are used the layout can be sensitive to the location of these joints. This makes it natural to consider extending the basic ground structure approach to include the

* Additional author for this chapter is Dr. Pedro Jesús *Martínez Castejón*.

Topology Design Methods for Structural Optimization.
DOI: http://dx.doi.org/10.1016/B978-0-08-100916-1.00002-7

optimization of the position of the joints. Such an approach of integrated geometry and topology can be found in the Refs. [7,17,18] using hierarchical methods.

Some of the disadvantages of the ground structure approach can be solved with a growth process, where joints and bars are added to a simple initial structure. The growth methods would start with a very simple topology and continually modify it by adding iteratively, joints and bars optimizing the joint geometry and bar sizes.

To automatically synthesize the topology of plane trusses with minimum inputs, Rule [19] proposed a technique based on the process of evolution founded in nature. A joint is introduced for every iteration at the mid-point of the largest bar, and is connected to the nearest joint by a newly added bar. The joint positions are then optimized before proceeding to the next iteration.

Mckeown [20] proposed a method that begins from the simplest possible geometry and proceeds by introducing joints, one at a time, optimizing both geometry and layout at each stage. The number and initial position of the new joints are fixed by the user.

Bojczuk and Mróz [21] proposed an algorithm that introduces new joints and bars into the optimal truss configuration. In order to minimize the cost function it assumes that the structure evolves and a bifurcation of the topology occurs with the generation of new joints. Their work considers two types of topology variation: (1) A new joint is generated at the center of an existing bar and is connected to an existing joint and (2) One existing joint separates into two joints with a new bar connecting them. In the first case, the stiffness of the added bar is the topological variable, whereas in the second it is the joint separation. Once a topology variation occurs, these parameters are optimized.

Kirsch [22] proposed a two-stage procedure, which consisted of both a reduction and expansion processes. Topology optimization is used to reduce the number of members and joints from a ground structure to generate an Initial Reduced Structure (IRS) with limited members and joints. The expansion process consists of successively adding members and joints to the IRS and carrying out a geometry and size optimization.

This chapter presents the growth method developed by Martínez et al. [23], for the optimal design of statically determinate plane truss structures using a sequential process of topology, followed by geometry and size optimization. The method grows a structure from nothing toward the optimum, rather than requiring an initial ground structure. It is applicable to structures subjected to a single load case with stress and size constraints.

2.2 THE GROWTH METHOD

The growth method combines sequentially the optimization of size, topology, and geometry of truss structures. The method has been applied to two-dimensional (2D) problems with a single load case and with stress and size constraints. The process given in Fig. 2.1 consists of five steps: (1) Domain

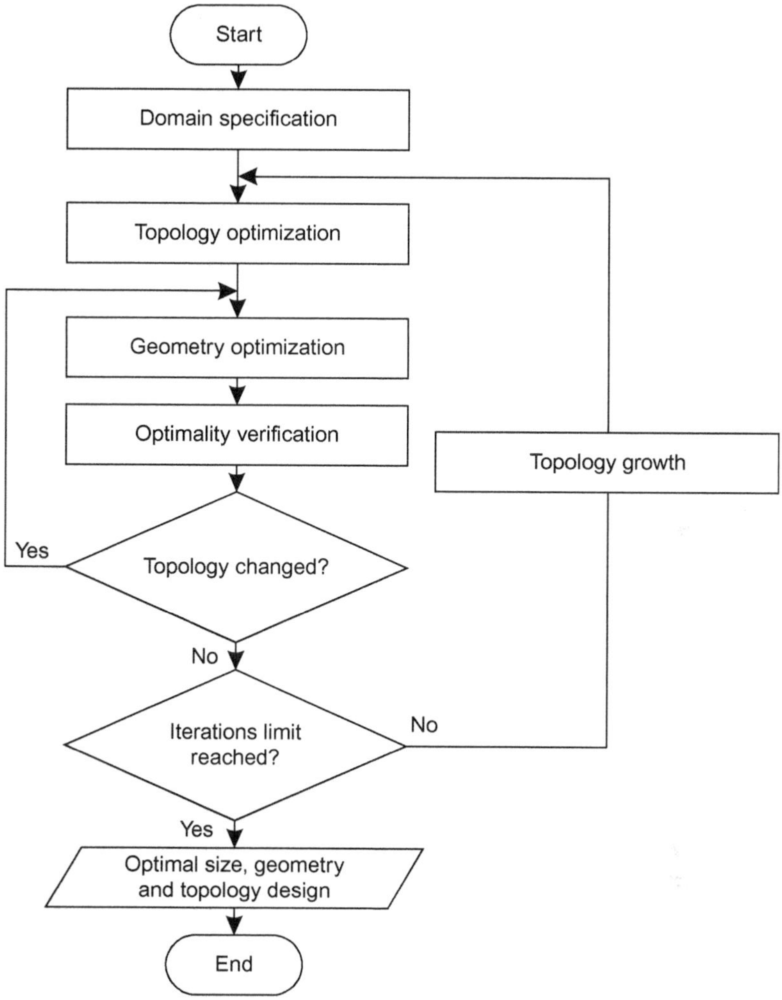

FIGURE 2.1 Flow chart showing the growth method.

specification; (2) topology and size optimization; (3) geometry optimization; (4) optimality verification; and (5) topology growth.

The process starts by specifying the location of the supports and loads together with their magnitude, the material information, and the maximum domain space which the resulting structure can occupy. The process is initiated by specifying a structural domain where all of the supports are connected to all of the applied loads. Topology and size optimization is then carried out. This is followed by geometry optimization, which modifies the position of the joints to improve the objective. The growth process is carried out next, which consists of systematically introducing a new joint and connecting it to neighbouring joints. This generates a number of structures, each with only a single new joint. In order to determine which joint lowers the objective, each new topology is optimized for both size and topology. The structure with the lowest objective is then selected as the new structure which produces the new design domain. The process is then repeated until no improvement is found on the objective or a limit on the number of iterations is reached. The following sections explain in detail each of the five steps in the process.

2.3 DOMAIN SPECIFICATION

For the growth process to start, an initial structure is required. This can either be generated automatically or by the user, the process consists of connecting each supporting joint with every applied load using a bar. Fig. 2.2 shows the initial domain for the Michell cantilever [24].

2.4 TOPOLOGY AND SIZE OPTIMIZATION

The method used for the topology optimization stage was that of Oberndorfer et al. [6]. It uses the ground structure approach with compliance as the criteria, which is equivalent to minimizing the mass for a single load case problem with stress as the constraint [25,26]. One thing to note is that the resulting optimal topology must be statically determinate [4,5,8], so any bar with a size at the lower limit that violates this constraint is automatically removed. Fig. 2.3 shows the optimal topology of the initially specified domain for the Michell cantilever of Fig. 2.2. The solid lines represent the bars, which produce the statically determinate optimal topology and the

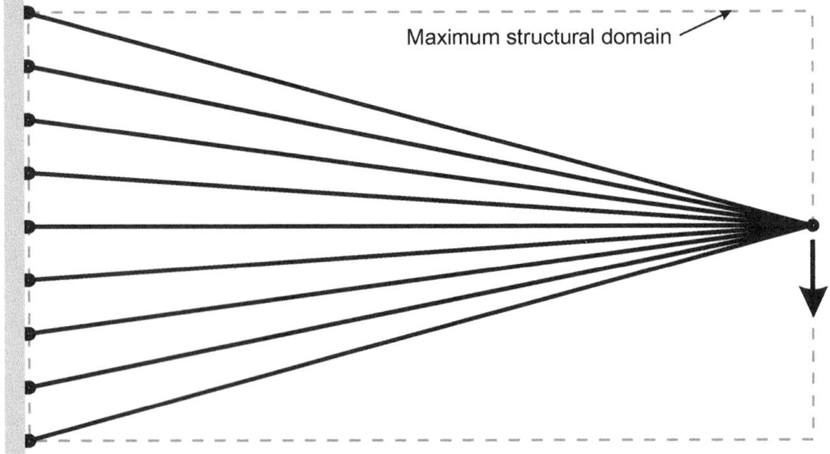

FIGURE 2.2 Initial domain for the Michell cantilever.

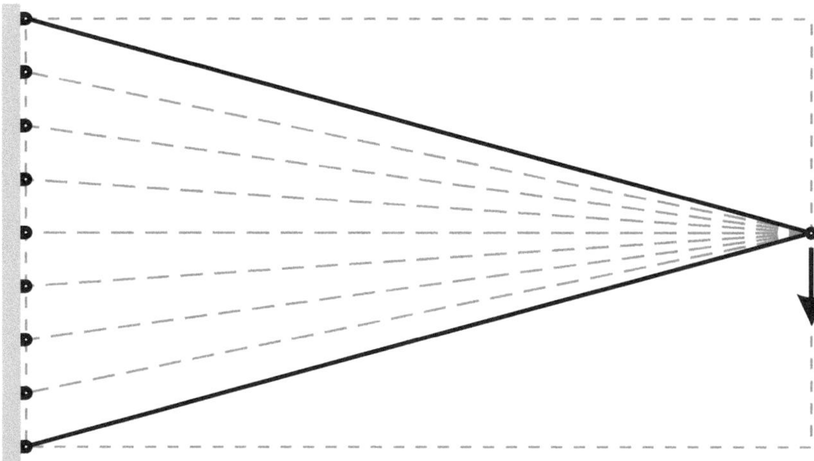

FIGURE 2.3 Optimal topology of the initial domain for the Michell cantilever.

dashed lines are those which are removed as they violate the constraint that the resulting structure must be statically determinate.

2.5 GEOMETRY OPTIMIZATION

The objective of the geometry optimization step is to minimize the mass of the statically determinate truss structure (Eq. 2.1), with the coordinates of the joints as the design variables subjected to their limits (Eq. 2.3). Stress constraints are not required as they are implicitly satisfied by Eq. (2.2). This resulted in a nonlinear geometrical optimization problem that can be solved using any unconstrained nonlinear optimization algorithm such as the Davidon-Fletcher-Powell (DFP) algorithm [27–29]. Whenever a side constraint limit is exceeded, i.e., the position of a joint moves outside the allowable design domain (Eq. 2.3), the problem becomes constrained. When this happens, a suitable algorithm can be used, such as the version of the Broyden-Fletcher-Goldfarb-Shanno nonlinear optimization algorithm [30] with limited memory and subject to simple bounds (L-BFGS-B) [31,32].

$$\text{Minimize:} \quad M(\mathbf{x}) = \sum_{e=1}^{m} \rho_e L_e A_e \tag{2.1}$$

$$\text{Subject to: } A_e = \begin{cases} 0 & \text{if } q_e = 0 \\ \dfrac{q_e}{\sigma_T} & \text{if } q_e > 0 \\ -\dfrac{q_e}{\sigma_T} & \text{if } q_e < 0 \end{cases} \tag{2.2}$$

$$\text{and} \quad x_i^L \leq x_i \leq x_i^U \quad i = 1, 2, \ldots, n \tag{2.3}$$

where m is the number of bars, $M(\mathbf{x})$ is the structural mass, ρ_e is the material density, L_e is the length, A_e is the cross-sectional area and q_e is the axial load of the eth bar; n is the number of design variables; $\mathbf{x}$ is the vector of the joints coordinates which are the design variables; x_i is the ith design variable with x_i^L the lower and x_i^U the upper bound of x_i; σ_T is the tensile; and σ_C is the compressive yield stresses.

The derivative of Eq. (2.1) is given by Eq. (2.4)

$$\frac{\partial M}{\partial x_i} = \sum_{e=1}^{m} \rho_e \left(\frac{\partial L_e}{\partial x_i} A_e + L_e \frac{\partial A_e}{\partial x_i} \right) \tag{2.4}$$

where the length and directional cosines for the eth bar are defined by Eqs. (2.5) and (2.6).

$$L_e = \sqrt{(x_e^k - x_e^j)^2 + (y_e^k - y_e^j)^2} \tag{2.5}$$

$$\cos\alpha = \frac{x_e^k - x_e^j}{L_e} \qquad \cos\beta = \frac{y_e^k - y_e^j}{L_e} \tag{2.6}$$

where j and k represents the two ends of the eth bar in 2D space and therefore

x_e^j and x_e^k are the x coordinates, and y_e^j and y_e^k are the y coordinates, respectively of the two ends of the eth bar.

The derivatives for the length and area terms with respect to the design variables then become Eqs. (2.7) and (2.8).

$$\frac{\partial L_e}{\partial x_i} = \begin{cases} 0 & \not\subset \ bar\ e \\ -\cos\alpha & \equiv x_e^j \\ \cos\alpha & \text{if } x_i \ \equiv x_e^k \\ -\cos\beta & \equiv y_e^j \\ \cos\beta & \equiv y_e^k \end{cases} \tag{2.7}$$

$$\frac{\partial A_e}{\partial x_i} = \frac{\partial A_e}{\partial q_e} \frac{\partial q_e}{\partial x_i} \tag{2.8}$$

The equation for equilibrium of the joints is given by Eq. (2.9)

$$\mathbf{Rq} = \mathbf{f} \tag{2.9}$$

where $\mathbf{R}$ is the matrix of directional cosines, $\mathbf{q}$ is the vector of axial forces, and $\mathbf{f}$ is the vector of external forces.

The necessary derivatives of the area with respect to the load carried by the eth bar and the loads with respect to the design variables are given by Eqs. (2.10), (2.11), and (2.12).

$$\frac{\partial A_e}{\partial q_e} = \begin{cases} 0 & \text{if } q_e = 0 \\ \dfrac{1}{\sigma_T} & \text{if } q_e > 0 \\ -\dfrac{1}{\sigma_T} & \text{if } q_e < 0 \end{cases} \tag{2.10}$$

$$\frac{\partial \mathbf{q}}{\partial x_i} = -\mathbf{R}^{-1}\frac{\partial \mathbf{R}}{\partial x_i}\mathbf{q} \tag{2.11}$$

$$\begin{array}{cc} \dfrac{\partial \cos\alpha}{dx_i} & \dfrac{\partial \cos\beta}{dx_i} \\ \Downarrow & \Downarrow \end{array}$$

$$\frac{\partial \mathbf{R}}{\partial x_i} = \left\{ \begin{array}{ccl} 0 & 0 & \not\subset \ bar\ e \\ -\dfrac{1-\cos^2\alpha}{L_e} & \dfrac{\cos\beta\cos\alpha}{L_e} & \equiv x_e^j \\ \dfrac{1-\cos^2\alpha}{L_e} & -\dfrac{\cos\beta\cos\alpha}{L_e} & \equiv x_e^k \\ \dfrac{\cos\alpha\cos\beta}{L_e} & -\dfrac{1-\cos^2\beta}{L_e} & \equiv y_e^j \\ -\dfrac{\cos\alpha\cos\beta}{L_e} & \dfrac{1-\cos^2\beta}{L_e} & \equiv y_e^k \end{array} \right. \quad \text{if } x_i \tag{2.12}$$

The flow chart of the geometry optimization process is given in Fig. 2.4, where the optimal value for each design variable (x_i) is represented by (x_i*).

2.6 OPTIMALITY VERIFICATION

As the locations of the joints are changed by the geometry optimization process, the location of the newly introduced joint may change considerably (usually at the early stages of the optimization). This change may give rise to the existence of a more optimal topology for the new geometry. For this reason the last added bar from the previous topology optimization is removed and new bars are connected from the last added joint to all other joints. A topology optimization is then carried out and if the resulting topology has a lower objective, it is retained and a new geometry optimization is carried out.

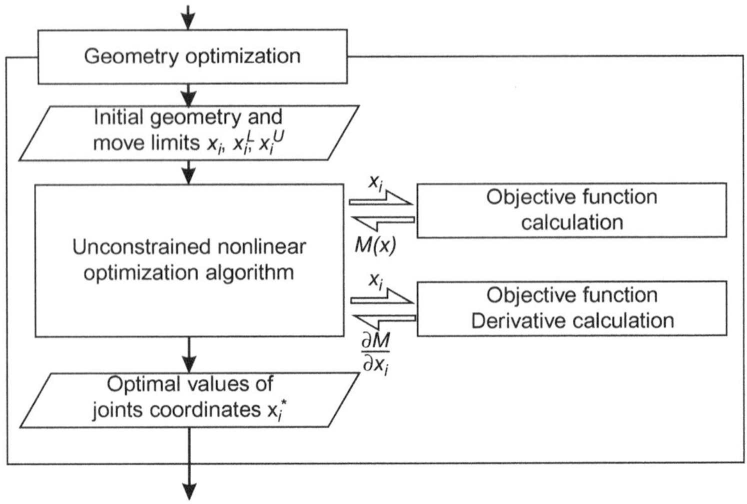

FIGURE 2.4 Flow chart of the geometry optimization process.

2.7 TOPOLOGY GROWTH

The growth process consists of specifying the location of a new joint, connecting it to neighbouring joints, and carrying out a topology optimization locally to determine which of these bars should remain.

A new joint can be introduced in one of the two ways: (1) By creating one where two bars cross each other or (2) by bisecting a bar. This guarantees improved orthogonality for the neighbouring joints.

For the case of intersecting bars, each of the two bars has coordinates (j_1, k_1) and (j_2, k_2), with gradients of their perpendiculars (m_{j1}, m_{k1}) and (m_{j2}, m_{k2}) respectively, Fig. 2.5. The location of the new joint is then given by the intersection of the two cubic polynomials or other selected splines.

For the case of bisecting a bar, the coordinates of its joints (j and k) and the gradients of their perpendiculars (m_j and m_k) are known, Fig. 2.6. The location of the new joint can then be calculated by bisecting a cubic polynomial (which is what was done in this work) or any other spline function fitted through these joints.

Once the position of the new joint is determined, bars are attached from here to all other neighbouring joints. This is followed by topology and size optimization to determine which of the added bars should remain in the structure and what its size should be, Fig. 2.7.

This process is repeated for every bar in the structure. The topology with the location of the new joint, which has the lowest objective is selected as the new structure. The entire process is then repeated until it converges or the iteration limit is reached.

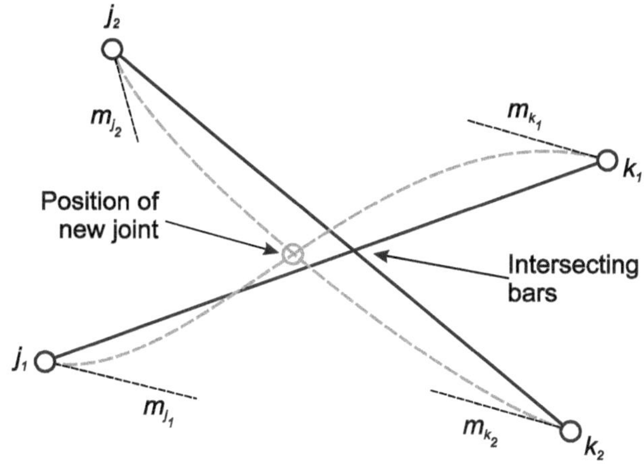

FIGURE 2.5 Location of the new joint for crossing bars.

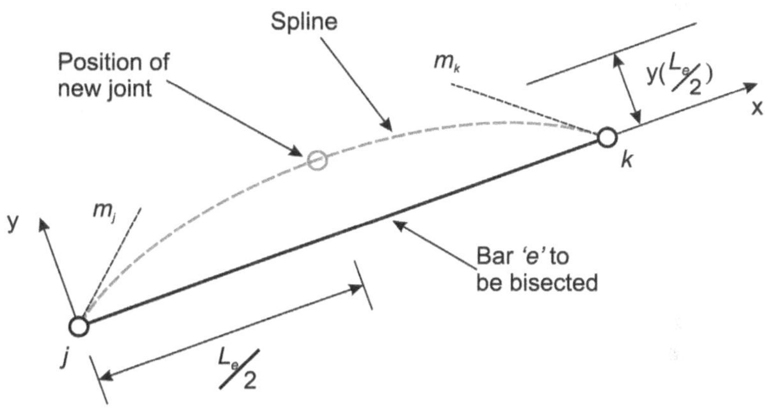

FIGURE 2.6 Location of the new joint.

2.8 PRACTICAL CRITERIA TO LIMIT THE NUMBER OF ADDED BARS TO NEW JOINTS

In the explanation of the growth method of Section 2.7, when a new joint is introduced, it is connected with bars to all other joints. However, a significant number of these bars are not practical. Two limiting criteria can be incorporated to reduce the number of added bars and to help speed up the topology optimization process. These were incorporated in the program included with this book.

2.8.1 Limiting the Number of Crossed Bars

When new bars are connected between the new joint and all other joints, each new bar should cross over no more than one of the existing bars, Fig. 2.8.

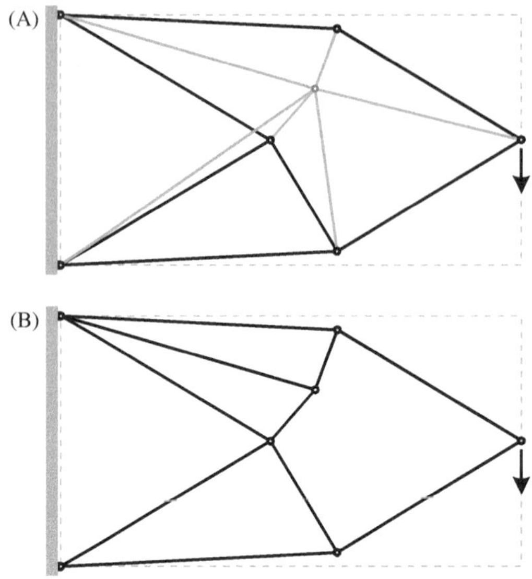

FIGURE 2.7 (A) Newly generated joint is connected to all other joints with bars. (B) After size and topology optimization leaving only one bar to make the remaining structure statically determinate.

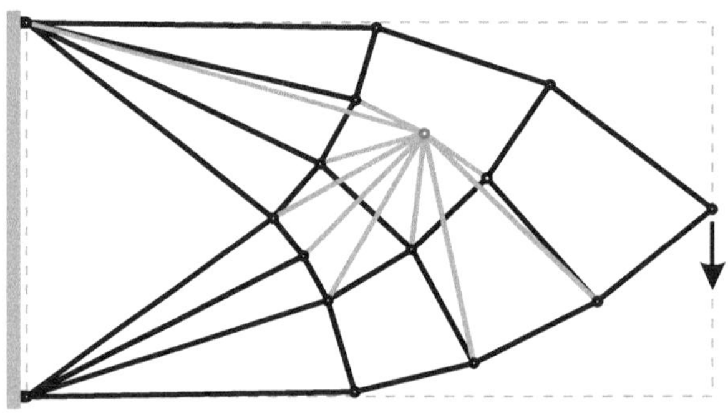

FIGURE 2.8 Michell cantilever showing the bars connected to a new joint but which cross over only one existing bar.

2.8.2 Using Orthogonality and Maximum Degree of Indeterminacy

The emerging topology must be statically determinate and the bars connecting each joint need to be as orthogonal to each other as possible. These principles can be applied to limit the number of added bars. A level of **I**ndeterminacy (***I***) given by Eq. (2.13) is specified, and the new bars are connected from the new joint to other adjacent joints where the angle of these new bars decrease in magnitude from 90 degrees until the number of bars is

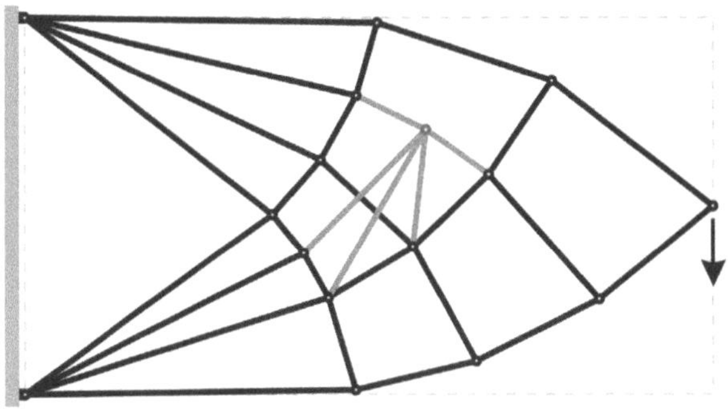

FIGURE 2.9 Michell cantilever with a level of indeterminacy of $I = 2$ and with newly connected bars as close to the perpendicular of the removed bar as possible.

equal to $(I + 1)$. For the Fig. 2.9, a value of $I = 2$ was used. Note that when $I = 0$, the structure is statically determinate.

$$I = M + R - 2j \tag{2.13}$$

where I is the Indeterminacy value, M is the number of bars, R is the number of reaction forces, and j is the number of joints in the structure.

REFERENCES

[1] A.G.M. Michell, The limit of economy of material in frame structures, Phil. Mag. 8 (6) (1904) 589–597.

[2] W. Prager, Optimal design of grillages, in: M. Save, W. Prager (Eds.), Structural Optimization, Plenum Press, New York, 1985, pp. 153–200.

[3] W. Prager, G.I.N. Rozvany, Optimization of structural geometry, in: A. Bednarek, L. Cesari (Eds.), Dynamical Systems, Academic Press, New York, 1977.

[4] W. Dorn, R. Gomory, H. Greenberg, Automatic design of optimal structures, J. Mecanique 3 (1964) 25–52.

[5] P. Fleron, The minimum weight of trusses, Bygnings statiske Meddelelser 35 (1964) 81–96.

[6] J.M. Oberndorfer, W. Achtziger, H.R.E.M. Hörnlein, Two approaches for truss topology optimization: a comparison for practical use, Struct. Optim 11 (3–4) (1996) 137–144.

[7] P. Pedersen, On the minimum mass layout of trusses. AGARD Conference Proceedings No. 36, Symposium on Structural Optimization, 1970, pp. 189–192.

[8] P. Pedersen, Topology optimization of three-dimensional trusses, in: M.P. Bendsøe, C.A. Mota Soares (Eds.), Topology Design of Structures, Kluwer, Dordrecht, 1993, pp. 19–30.

[9] O. da Silva Smith, An interactive system for truss topology design, Adv. Eng. Softw. 27 (1–2) (1996) 167–178.

[10] W. Achtziger, M.P. Bendsøe, A. Ben-Tal, J. Zowe, Equivalent based formulations for maximum strength truss topology design, Impact Comp. Sci. Eng. 4 (4) (1992) 315–345.

[11] A. Ben-Tal, M.P. Bendsøe, A new method for optimal truss topology design, SIAM J. Optim. 3 (1993) 322–355.

[12] M.P. Bendsøe, A. Ben-Tal, J. Zowe, Optimization methods for truss geometry and topology design, Struct. Optim 7 (3) (1994) 141–158.

[13] W. Achziger, Truss topology optimization including bar properties different for tension and compression, Struct. Optim 12 (1) (1996) 63–73.
[14] M. Zhou, G.I.N. Rozvany, DCOC: an optimality criteria method for large systems, Part I: Theory. Struct. Optim 5 (1992) 12–25.
[15] M. Zhou, G.I.N. Rozvany, DCOC: an optimality criteria method for large systems, Part II: algorithm. Struct. Optim 6 (1993) 250–262.
[16] M. Zhou, G.I.N. Rozvany, An improved approximation technique for the DCOC method of sizing optimization, Comp. Struct 60 (5) (1996) 763–769.
[17] F. Nishino, R. Duggal, Shape optimum design of trusses under multiple loading, Int. J. Solids Struct. 26 (1990) 17–27.
[18] A. Ben-Tal, M. Kocvara, J. Zowe, Two non-smooth approaches to simultaneous geometry and topology design of trusses, in: M.P. Bendsøe, C.A. Mota Soares (Eds.), Topology Design of Structures, Kluwer, Dordrecht, 1993, pp. 31–42.
[19] W.K. Rule, Automatic truss design by optimized growth, J. Struct. Eng. 120 (10) (1994) 3063–3070.
[20] J.J. McKeown, Growing optimal pin-jointed frames, Struct. Optim 15 (1998) 92–100.
[21] D. Bojczuk, Z. Mróz, Optimal topology and configuration design of trusses with stress and buckling constraints, Struct. Optim. 17 (1999) 25–35.
[22] U. Kirsch, Integration of reduction and expansion processes in layout optimization, Struct. Optim. 11 (1–2) (1996) 13–18.
[23] P. Martínez, P. Martí, O.M. Querin, Growth method for size, topology and geometry optimization of truss structures, Struct. Multidiscipl. Optim. 33 (Number 1) (2007) 13–26.
[24] G.I.N. Rozvany, Exact analytical solutions for some popular benchmark problems in topology optimization, Struct. Optim 15 (1998) 42–48.
[25] W.S. Hemp, Optimum Structures, Clarendon Press, Oxford, 1973.
[26] H.L. Cox, The Design of Structures for Least Weight, Pergamon Press, Oxford, 1965.
[27] W.C. Davidon, Variable metric method for minimization, SIAM J. Optim. 1 (1) (1991) 1–17.
[28] R. Fletcher, M.J.D. Powell, A rapidly convergent descent method for minimization, Comp. J. 6 (2) (1963) 163–168.
[29] Davidon, W.C. 1959: Variable metric method for minimization. A.E.C. Research and Development Report, ANL-5990.
[30] J.E. Dennis, J.J. More, Quasi-Newton methods, motivation and theory, SIAM Rev. 19 (1977) 46–89.
[31] J.L. Morales, J. Nocedal, Remark on "algorithm 778: L-BFGS-B: Fortran subroutines for large-scale bound constrained optimization", ACM Trans. Math. Softw. 38 (1) (2011) Article 7.
[32] C. Zhu, R.H. Byrd, P. Lu, J. Nocedal, Algorithm 778: L-BFGS-B: FORTRAN subroutines for large-scale bound constrained optimization, ACM Trans. Math. Softw. 23 (4) (1997) 550–560.

Chapter 3

Discrete Method of Structural Optimization*

3.1 INTRODUCTION

Numerical finite element (FE) methods were first applied to structural topology optimization by Bendsøe and Kikuchi [1], who used a homogenization method for ground elements with square or rectangular holes. This method was gradually replaced by the Solid Isotropic Microstructure with Penalization (SIMP) method, first introduced by Bendsøe [2], but independently developed by Zhou and Rozvany [3]. In SIMP, the elements are considered homogeneous and isotropic, and the design variables consist of the density (or thickness) of each element, whose intermediate values are penalized. Material interpolation schemes corresponding to SIMP were shown by Bendsøe and Sigmund [4].

Structural topology optimization is discussed in these excellent articles [5–9].

Other methods were developed in parallel to SIMP, such as the Evolutionary Structural Optimization (ESO) [10], Additive ESO (AESO) [11] and Bi-directional ESO (BESO) methods [12,13], Reverse Adaptivity [14], Metamorphic Development [15], Genetic algorithms [16] or the Level Set method [17–20].

In the ESO method [10,21], the structure evolves toward an optimum after inefficient elements are gradually removed. The additive version of this method, AESO [11,12] enabled the admission of new elements around highly stressed elements present in the structure. These two strategies were then combined to produce the bidirectional version of this method, BESO [12,13]. The BESO method allowed elements to be both added and removed from the design domain.

Rozvany and Querin [22–24] proposed the Sequential Element Rejection and Admission (SERA) method as the alternative to any ESO/BESO method, which addresses some problems with the ESO methods which were highlighted by various authors [25–27]. The SERA method is also bidirectional in nature but works with two separate criteria for adding and removing

* Additional author for this chapter is Prof. Dr.-Ing. Dr.-Habil. George I. N. Rozvany.

Topology Design Methods for Structural Optimization.
DOI: http://dx.doi.org/10.1016/B978-0-08-100916-1.00003-9

stress-based material from the design domain. The method was developed for stress-based design [23,24,28–31] and for compliant mechanisms design [32–35], where the versatility and robustness of the method was clearly demonstrated.

This chapter provides a detailed explanation of the SERA method for structural topology optimization and for compliant mechanisms. The procedure is presented in a generalized way for any problem formulation and specified for four different driving criteria: (1) fully stressed design, (2) minimum compliance design, (3) multiple load conditions, and (4) multiple support conditions. This chapter presents the published work on the development of this method [24,26,28,29,31–35].

3.2 THE SEQUENTIAL ELEMENT REJECTION AND ADMISSION (SERA) ALGORITHM

The SERA method is bidirectional in nature and considers two separate material domains: (1) the "*Real*" material domain and (2) the "*Virtual*" material domain, where the elements have an almost negligible material stiffness [23,24,30]. Two separate criteria for the rejection and admission of elements allow material to be introduced and removed by changing their status from "*virtual*" to "*real*" and vice versa (Fig. 3.1). The final topology consists of all the "*real*" material present at the end of the optimization.

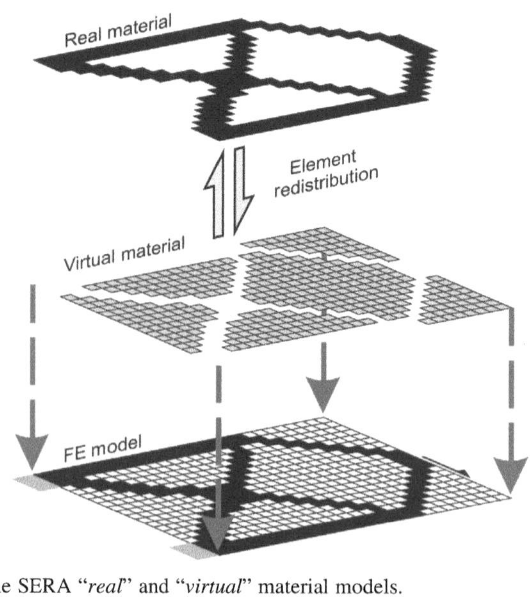

FIGURE 3.1 The SERA "*real*" and "*virtual*" material models.

The driving criterion can consist of one of the following two approaches:

1. The behavioral constraints (i.e., limits on the values or response parameters, or their weighted combinations) are given, and the volume, mass, or cost of the structure is minimized.
2. A constraint on the cost (i.e., volume, volume fraction, or mass) is given, and a value (i.e., maximum or minimum) of the designated response parameter is minimized or maximized.

Approach (1) is necessary for multiconstraint problems and is used more often in engineering practice, whereas approach (2) is used by the SERA method. In both approaches, a so-called "criterion value" is calculated for each element of both the real and virtual material domains. In the original formulation of the SERA method for stressed-based criteria, an element of the real material domain is moved to the virtual domain if its criteria value is low, whereas an element of the virtual material domain is moved to the real domain if its criteria value is high. The criterion value can be any response parameter for the considered element, or some convex combination of such parameters.

The general SERA algorithm for structural optimization problems consists of the following 13 steps and is also depicted in the flow chart of Fig. 3.2.

1. Define the design problem: This consists of the maximum domain that can be occupied by the structure, the finite element mesh, all boundary conditions and loads, the real and virtual material properties, and the objective function (Section 3.3).
2. Specify the SERA parameters: These consist of the limit volume fraction (V^{Lim}), the total number of iterations (N_{Tot}) to reach this limit volume or the progression rate (PR), and the smoothing ratio (SR) (Section 3.4).
3. Specify the initial design domain and assign the "*real*" and "*virtual*" material domains (Section 3.5).
4. Calculate the volume fraction to be redistributed in the current (ith) iteration ($\Delta V(i)$), which consists of the material domains to be added ($\Delta V_{\mathrm{Add}}(i)$) and removed ($\Delta V_{\mathrm{Rem}}(i)$) (Section 3.6).
5. Carry out an analysis of the maximum structural domain using the Finite Element Method (FEM) (Section 3.7).
6. Calculate the criterion value for each element (Cv_e) (Section 3.8).
7. Apply mesh independent filtering (Section 3.9).
8. Separate the element criterion values for the "*real*" (Cv_R) and "*virtual*" (Cv_V) material domains (Section 3.8).
9. Calculate the threshold values for the "*real*" (Cv_R^{thr}) and "*virtual*" (Cv_V^{thr}) material domains. These values are used to determine which elements are removed and which are introduced into the real domain (Section 3.8).
10. Redistribute elements from the "*real*" to "*virtual*" domain and vice versa (Section 3.6).

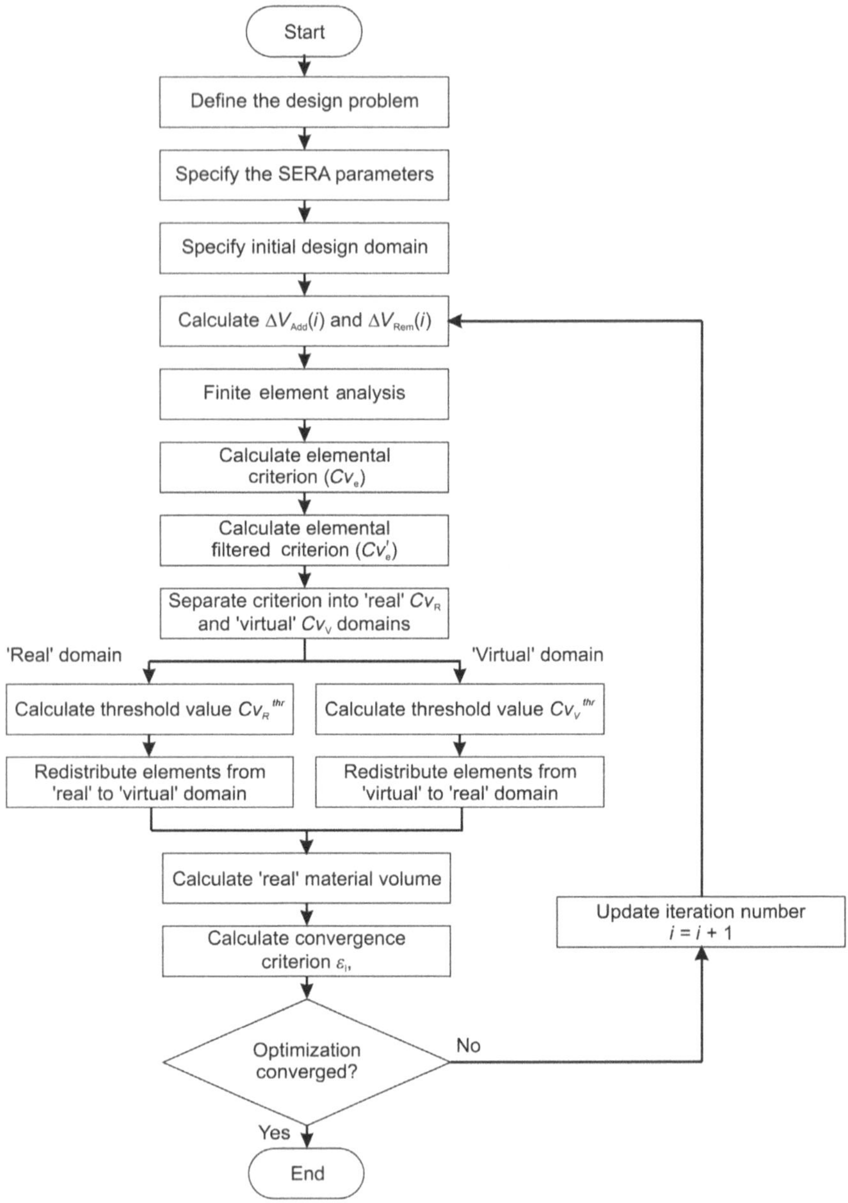

FIGURE 3.2 Flow chart of the SERA method.

11. Calculate the "*real*" material volume in the domain.
12. Calculate the current iteration convergence criterion value (ε_i) (Section 3.10).
13. Repeat steps (4)–(12) until the limit volume fraction (V^{Lim}) is reached and the convergence criterion for the objective function is satisfied

(Section 3.10). The final topology consists of only the "*real*" material in the design domain.

3.3 DEFINITION OF THE OBJECTIVE FUNCTION

When a structure is developed, the designer may want all of its components to be fully utilized or alternatively a requirement might be for it to have the least deflection possible. In terms of structural optimization, these two conditions correspond to the structure being fully stressed or for its compliance to be minimized. There may of course, be other driving factors such as natural frequency or structural stability (buckling), but these are not covered in this book. Structures, of course, experience different loading condition, or can have different active supports at different times, thus requiring multiple cases to be considered when carrying out their optimization.

Compliant mechanisms produce one or several displacements in specific directions due to one or several input motions elsewhere within its domain. Fig. 3.3 shows a compliant mechanism domain Ω subjected to a force F_{in} at the input port P_{in} which is supposed to produce an output displacement Δ_{out} at the output port P_{out}. Their optimization requires the maximization of the output displacements or force due to the work done to the mechanism. This is expressed as the maximization of the Mutual Potential Energy (MPE).

The SERA method has been applied to four types of problems: (1) stress [28,29,31]; (2) compliance [35]; (3) multiple criteria [35]; and (4) MPE [34,35]. The corresponding objective functions are explained in the following sections.

3.3.1 Stress-Based Objective Function

The stress-based objective function consists of the minimization of the maximum von Mises stress (Eq. 3.1) subject to a target volume fraction (Eq. 3.2). This was shown by Brodie [31] to give as close as possible to a fully stressed design.

$$\text{minimize} \quad \sigma_{vM_{\text{Max}}} \tag{3.1}$$

$$\text{subject to} \quad V_{(\rho_e)} = \sum_{e=1}^{N} \rho_e \frac{V_e}{V_{\text{Tot}}} \leq V^{\text{Lim}}; \; e = 1, \ldots, N; \; \rho_e = \{\rho_{\min}, 1\} \tag{3.2}$$

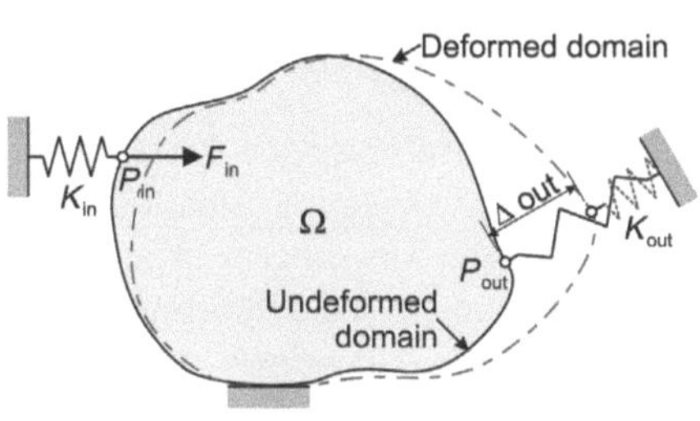

FIGURE 3.3 Compliant mechanism problem definition.

where $\sigma_{vM_{Max}}$ is the maximum von Mises stress in the structure which for a general continuum domain is calculated using (3.3); ρ_e is the "*virtual*" density and V_e is the volume of the eth finite element; V_{Tot} is the total volume of the maximum domain that can be occupied by the structure; V^{Lim} is the desired target volume fraction specified by the user; N is the number of finite elements in the mesh and $\rho_{\min}$ is the minimum virtual elemental density, the value used in the work presented in this book is $\rho_{\min} = 10^{-4}$.

$$\sigma_{vM} = \sqrt{\frac{1}{2}\left[(\sigma_1-\sigma_2)^2 + (\sigma_2-\sigma_3)^2 + (\sigma_1-\sigma_3)^2\right]} \tag{3.3}$$

where σ_1, σ_2, and σ_3 are the principal stresses.

3.3.2 Compliant-Based Objective Function

The compliance-based objective function consists of the minimization of the structural compliance C (Eq. 3.4), subject to a target volume fraction (Eq. 3.2). This minimization of compliance is equivalent to the maximization of the rigidity of the structure.

$$\text{minimize} \quad C = F^T u, \;\; Ku = F \tag{3.4}$$

$$\text{subject to: } V_{(\rho_e)} = \sum_{e=1}^{N} \rho_e \frac{V_e}{V_{\mathrm{Tot}}} \leq V^{\mathrm{Lim}}; \;\; e = 1,\ldots,N; \;\; \rho_e = \{\rho_{\min}, 1\} \tag{3.2}$$

where K is the global stiffness matrix of the structure, F is the nodal force vector, and u is the displacement field.

3.3.3 Multiple-Criteria Objective Function

Optimization problems with multiple criteria, such as multiple loading or active support conditions, can be addressed by combining the different objective functions using the weighting factor approach of Eq. (3.5).

$$\begin{gathered} \mathrm{OF_{MC}} = \sum_{k=1}^{K} \mathrm{OF}_k^{*} \omega_k, \\ \text{where } \sum_{k=1}^{K} \omega_k = 1 \end{gathered} \tag{3.5}$$

where $\mathrm{OF_{MC}}$ is the multiple-criteria objective function, K is the total number of criteria, OF_k^{*} is the normalized objective function for the kth criteria, and ω_k is the weighting factor for the kth criteria. The values of the individual weighting factors for each criteria can be determined in one of two ways:

1. They can be specified by the designer; or

2. They can be calculated using Eq. (3.6), which consists of calculating the proportion of the structural compliance of the kth criteria to that of total structural compliance of all criteria combined.

$$\omega_k = \frac{C_k}{\sum_{k=1}^{K} C_k} \tag{3.6}$$

3.3.4 Mutual Potential Energy Objective Function

The MPE objective function consists of the maximization of the MPE (3.7), subject to a target volume fraction (Eq. 3.2).

$$\text{minimize} \quad \text{MPE} \tag{3.7}$$

$$\text{subject to} \quad V_{(\rho_e)} = \sum_{e=1}^{N} \rho_e \frac{V_e}{V_{\text{Tot}}} \leq V^{\text{Lim}}; \quad e = 1, \ldots, N; \quad \rho_e = \{\rho_{\min}, 1\} \tag{3.2}$$

The MPE (Eq. 3.8) [36] is the deformation at a prescribed output port in a specified direction. To calculate the MPE, the displacements due to two load cases are required: (1) the input force case, where the input force F_{in} is applied to the input port P_{in}, defined with the subscript 1 in (Eqs. 3.8 and 3.9) and Fig. 3.4A; (2) the pseudo-force case, where a force of unit (1) magnitude is applied at the output port P_{out} in the direction of the desired displacement, defined with the subscript 2 in (Eqs. 3.8 and 3.10) and Fig. 3.4B.

$$\text{MPE} = u_2^T K u_1 \tag{3.8}$$

$$K u_1 = F_1 \tag{3.9}$$

$$K u_2 = F_2 \tag{3.10}$$

where $\boldsymbol{K}$ is the global stiffness matrix of the structure; $\boldsymbol{F}_1$ is the nodal force vector which contains the input force F_{in}; $\boldsymbol{F}_2$ is the nodal force vector which contains the unit output force F_{out}; and $\boldsymbol{u}_1$, $\boldsymbol{u}_2$ are the displacement fields due to each load case.

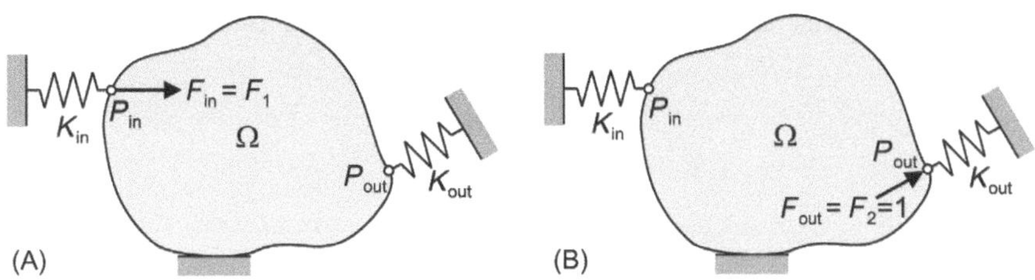

FIGURE 3.4 Representation of the load cases: (A) Case 1: input force; (B) Case 2: pseudo-force.

For the case where the compliant mechanism has multiple inputs and multiple outputs, the MPE is defined by Eq. (3.11).

$$\begin{aligned} \mathrm{MPE} &= \sum_{i=1}^{n}\sum_{j=1}^{m}\omega_{i,j}\mathrm{MPE}_{i,j} \\ \text{where: } &\sum_{i=1}^{n}\sum_{j=1}^{m}\omega_{i,j} = 1 \end{aligned} \tag{3.11}$$

$$\mathrm{MPE}_{i,j} = u_j^T K u_i \tag{3.12}$$

$$K u_i = F_i \tag{3.13}$$

$$K u_j = F_j \tag{3.14}$$

where $\omega_{i,j}$ is the weighting factor for each of the input/output combinations and as for the multiple-criteria case, can either be specified by the designer or be calculated using Eq. (3.6); $\boldsymbol{K}$ is the global stiffness matrix of the structure; $\boldsymbol{F}_i$ is the nodal applied force vector of the ith input case applied at the ith input ports P_i to produce the ith output case. $\boldsymbol{F}_j$ is the nodal unit force vector of the jth output case applied at the output port P_j; and $\boldsymbol{u}_i$, $\boldsymbol{u}_j$ are the displacement fields due to each of the above mentioned load cases.

3.4 THE SERA PARAMETERS

There are seven parameters which drive the SERA method, these are the: (1) limit volume fraction (V^{Lim}); (2) total number of iterations (N_{Tot}); (3) PR; (4) SR; (5) material redistribution ratio (β); (6) filter radius ($r_{\min}$); and (7) convergence limit ($\varepsilon_{\mathrm{Lim}}$).

3.4.1 The Limit Volume Fraction

The limit volume fraction (V^{Lim}) consists of the ratio of the desired final volume of the structure (V_{Max}) divided by the total volume of the maximum domain (V_{Tot}) that the structure can occupy, its value is calculated using Eq. (3.15).

$$V^{\mathrm{Lim}} = \frac{V_{\mathrm{Max}}}{V_{\mathrm{Tot}}} \tag{3.15}$$

3.4.2 Controlling the Rate of Material Admission and Removal

There are two ways to control how SERA removes and introduces material into the design domain:

1. Allocate the value of the PR and allow SERA to converge to the limit volume ratio by itself but with no control on how many iterations it will take before the material redistribution phase. Typical values of PR which the authors have used are in the range of 0.001–0.05. If this value is too low, it may take a long time to reach the optimum, but if it is too large, although reaching the desired volume relatively fast, the emerging topology may not be the optimum.
2. Specify the total number of iterations (N_{Tot}) and using Eq. (3.16) determine the PR required.

$$PR = \frac{|V_{Max} - V_{Init}|}{V_{Tot} \times N_{Tot}} \tag{3.16}$$

where V_{Init} is the initial or starting design domain volume. The reason for the absolute value sign is because the initial design domain volume can be less than the final maximum design domain volume required.

3.4.3 The Smoothing Ratio

At each iteration of the SERA method, material is both introduced and removed from the design domain. However, depending on the initial design domain (Section 3.5); there will either be a net increase or a net decrease of material. In order to control the level of excess material either added or removed, the SR is used. Typical values used in the examples presented in this book range between 1.2 and 1.4.

3.4.4 The Material Redistribution Fraction

Once the target volume is reached, due to the nature of SERA, there needs to be a settling stage where a very small amount of material is redistributed within the entire design domain. This means that exactly the same amount of material is added and removed until the criterion distribution converges. This is controlled by the material redistribution ratio (β), with typical values ranging between 0.001 and 0.005. This means that, somewhere between 0.1% and 0.5% of the volume domain is redistributed.

3.4.5 The Filter Radius

The filter radius (r_{min}) is used after the elemental criterion is calculated in order to suppress the checkerboard effect in the optimized solution. The filter modifies the elemental criterion of a specific element based on a weighted average of the criteria in a fixed neighborhood, [37]. The filter radius is specified by the user and needs to be at least larger than the size of the smallest FE use, in order to have an effect. The size of the filter radius can be used to control the minimum size of the emerging features in the design domain.

3.4.6 Convergence Limit

One of the mechanisms to control when the topology optimization process has finished is in the conference of the objective function optimal value. Typical values for the convergence limit (ε_{Lim}) are between 0.001 and 0.01. If the convergence value (ε_i) is less than the convergence limit (ε_{Lim}), then the optimization process is terminated.

3.5 THE INITIAL DESIGN DOMAIN

The initial design domain can consist of any of three different combinations of "*real*" and "*virtual*" material domains, these are:

1. A full design domain where all of the material is "*real*" material;
2. A void design domain, where all of the material is "*virtual*" material; or
3. With any amount of material present in the domain, this means a combination of "*real*" and "*virtual*" material.

For any of these cases, the material present in the domain is assigned the "*real*" material properties and material not present (or void) in the domain is assigned the "*virtual*" material properties. The method converges toward the optimum topology regardless of the initial design domain. However, the initial design domain determines the number of iterations needed to achieve that optimum.

For the case of structural designs, a full initial design domain (all "*real*" material) is the most efficient starting domain, producing the fastest converging solution to an optimum, [35]. For compliant mechanisms, a void initial design domain (all "*virtual*" material) is the most efficient starting domain, as it requires less number of iterations to reach the optimum than starting with different amounts of '*real material*' in the domain [33].

3.6 THE VOLUME FRACTION TO BE REDISTRIBUTED

The process of material redistribution consists of two stages, these are:

1. Different amounts of material are rearranged between the "*real*" and "*virtual*" design domain at each iteration, until the limit volume fraction (V^{Lim}) is reached (Section 3.6.1).
2. The material redistribution stage is only started once the limit volume fraction (V^{Lim}) is reached. This stage consists of a fixed volume of material which is both added and removed simultaneously from the design domain, until the problem converges (Section 3.6.2).

3.6.1 Determine the Volume Fraction to be Rearranged

The volume fraction to be rearranged depends on the starting design domain and on the iteration number. Two starting domain cases exist:

1. When the design starts with a volume fraction higher than the limit volume fraction or a full domain, ($V_F(0)$) or
2. When the design starts with a volume fraction lower than the limit volume fraction or from a void domain ($V_V(0)$).

For case (1), the target volume fraction ($V_F(i)$) is calculated using Eq. (3.17), and for case (b), the target volume fraction ($V_V(i)$) is calculated using Eq. (3.18). The total amount of material to be rearranged (both added and removed) in the ith iteration is then given by Eq. (3.19). This value is then separated into the volume fraction to be added ($\Delta V_{\text{Add}}(i)$) and the volume fraction to be removed ($\Delta V_{\text{Rem}}(i)$). For case (1), these terms are given by Eqs. (3.20) and (3.21), and for case (2) they are given by Eqs. (3.22) and (3.23).

$$V_F(i) = \max\left\{(V_F(i-1) \times (1 - PR)),\ V^{\text{Lim}}\right\} \tag{3.17}$$

$$V_V(i) = \min\left\{(V_V(i-1) + PR),\ V^{\text{Lim}}\right\} \tag{3.18}$$

$$\Delta V(i) = \left|V_{F\text{ or }V}(i) - V_{F\text{ or }V}(i-1)\right| \tag{3.19}$$

$$\Delta V_{\text{Add}\ F}(i) = \Delta V(i) \times (\text{SR} - 1) \tag{3.20}$$

$$\Delta V_{\text{Rem}\ F}(i) = \Delta V(i) \times \text{SR} \tag{3.21}$$

$$\Delta V_{\text{Add}\ V}(i) = \Delta V(i) \times \text{SR} \tag{3.22}$$

$$\Delta V_{\text{Rem}\ V}(i) = \Delta V(i) \times (\text{SR} - 1) \tag{3.23}$$

A graphical representation of the removal and addition of elements at each iteration for case (1) is given in Fig. 3.5 and for case (2) in Fig. 3.6. In both cases, each iteration consists of two substeps which both add and remove material from the design domain. The difference depends on the amount of material

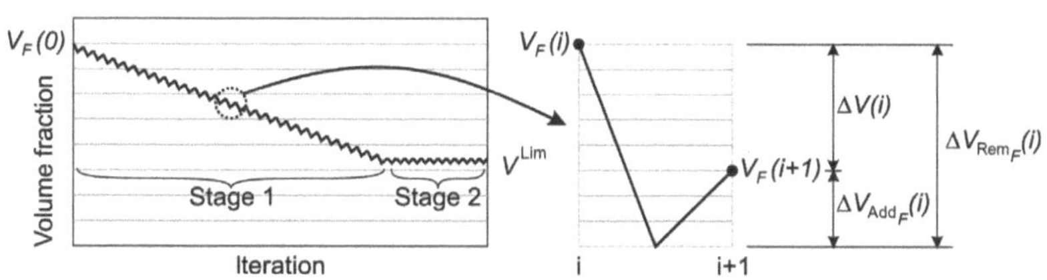

FIGURE 3.5 The material removal and addition process for Stage 1 starting from a domain with a higher volume fraction ($V_F(0)$) than the final domain (V^{Lim}).

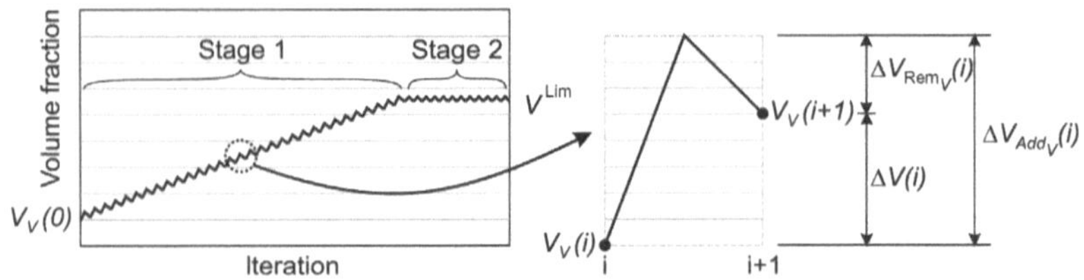

FIGURE 3.6 The material removal and addition process for Stage 1 starting from a domain with a lower volume fraction ($V_V(0)$) than the final domain (V^{Lim}).

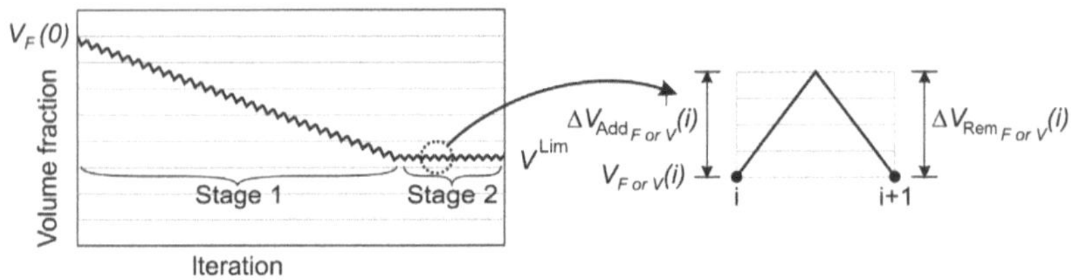

FIGURE 3.7 Material redistribution, which is Stage 2 of the material removal and addition process.

to be added or removed so that the volume fraction in that iteration decreases for case (1), Fig. 3.5, or increases for case (b), Fig. 3.6.

3.6.2 Material Redistribution

The process of material redistribution consists of both adding and removing the same amount of material from the design domain, Eq. (3.24) until the criterion distribution converges. This volume fraction to be redistributed is given by Eq. (3.24), where β is the material redistribution fraction. Typical values range between 0.001 and 0.005 which correspond to between 0.1% and 0.5% of the total domain volume. A graphical representation of the redistribution process during the ith iteration is shown in Fig. 3.7.

$$\Delta V_{\text{Add}_{F \text{ or } V}}(i) = \Delta V_{\text{Rem}_{F \text{ or } V}}(i) = \beta \times V^{\text{Lim}} \tag{3.24}$$

3.7 THE FINITE ELEMENT ANALYSIS

In order to analyze a structure using the FEM, the global stiffness matrix of the structure needs to be calculated, which consists of the sum of the stiffness matrices of each finite element (FE). However, unlike standard FEA, for this work, each elemental stiffness matrix is multiplied by the "*virtual*" elemental density (ρ_e), to give Eq. (3.25).

$$K = \sum_{e=1}^{N} \rho_e k_e, \quad e = 1, \ldots, N; \quad \rho_e = \{\rho_{\min}, 1\} \tag{3.25}$$

where K is the global stiffness matrix, ρ_e is the virtual elemental density for the eth FE; k_e is the stiffness matrix for the eth FE and N is the number of FEs in the mesh.

For all of the applications of the SERA method presented in this book, the FEA of the structures is straight forward. Structural problems require a single FEA for a single load condition or multiple FEAs when multiple load or support conditions are applied. Compliant mechanisms, however, require two load conditions to be analyzed for every input–output combination. These consist of the real input load applied at the input port and of a unit load applied at the corresponding output port.

3.8 THE ELEMENTAL CRITERION VALUE

The type of elements criterion value (Cv_e) or response parameter required by the SERA method depends on a combination of objective function used and the type of topology optimization sought (structural or compliant). Four elemental criteria have been implemented into SERA; these are: (1) stress for structural problems; (2) compliance for structural problems; (3) multiple loadings/supports for structural problems; and (4) single input–output compliant. These are explained in the following sections.

3.8.1 Elemental Criterion for a Fully Stressed Design

The elemental criterion (Cv_e) required to produce a fully stressed design is the von Mises stress (σ_{vM}). For a three-dimensional continuum domain, the elemental von Mises stress is given by Eq. (3.26). As the objective is to minimize the maximum von Mises stress and therefore produce a fully stressed design, the "*real*" elements with the lower elemental criterion are the ones removed and the "*virtual*" elements with the higher elemental criterion are the ones added, Fig. 3.8. The real and virtual elements are ordered in increasing elemental criterion value, as shown in Fig. 3.8. In order to determine the required volume fractions to be added (ΔV_{Add}) and removed (ΔV_{Rem}), the "*real*" and "*virtual*" material domains threshold values (Cv_R^{thr} and Cv_V^{thr}) have to be calculated.

$$Cv_e = \sigma_{vM_e} = \sqrt{\frac{1}{2}\left[(\sigma_{1_e} - \sigma_{2_e})^2 + (\sigma_{2_e} - \sigma_{3_e})^2 + (\sigma_{1_e} - \sigma_{3_e})^2\right]} \tag{3.26}$$

where σ_{1_e}, σ_{2_e}, and σ_{3_e} are the principal stresses, and σ_{vM_e} is the von Mises stress for the eth element.

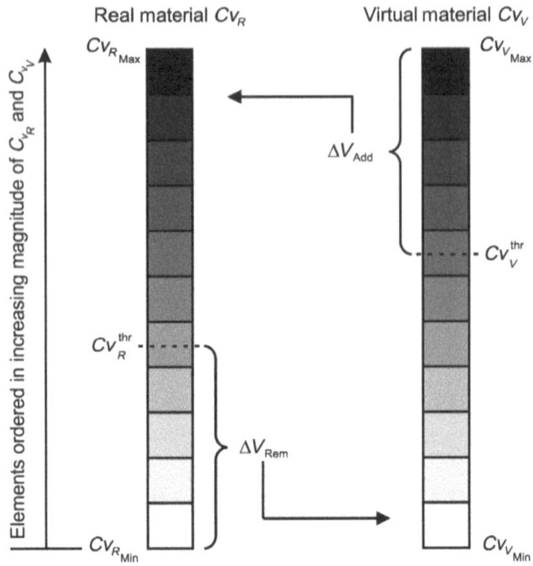

FIGURE 3.8 Ordered "*real*" and "*virtual*" material criterion domains for a fully stressed design case, showing the criterion threshold and the material volumes to be rearranged.

3.8.2 Elemental Criterion for Minimum Compliance

The elemental criterion (Cv_e) required for the structural problem of minimum compliance is given by Eq. (3.27) [10]. It provides information on how sensitive the compliance is to small changes in the design variables.

$$Cv_e = \alpha_e = u_e^T \Delta k_e u_e \tag{3.27}$$

where α_e is the sensitivity number of the eth element; u_e is the displacement vector of the eth element due to the applied load vector $\{F\}$ and Δk_e is the variation of the eth element elemental stiffness matrix between two iterations (Eq. 3.28).

$$\Delta k_e = k_e(i) - k_e(i-1) \tag{3.28}$$

where $k_e(i-1)$ and $k_e(i)$ are the elemental stiffness matrices for the eth FE in the (ith $-$ 1) and (ith) iterations, respectively.

If an element is removed in the ith iteration, then the stiffness matrix changes from having a value of $k_e(i) \approx 0$ to that of $k_e(i-1) = k_e$. So the variation of the elemental stiffness matrix using Eq. (3.27) becomes $\Delta k_e = -k_e$. But if an element is added in the ith iteration, then its stiffness matrix changes from having the value $k_e(i-1) \approx 0$ to that of $k_e(i) = k_e$. So the variation of the elemental stiffness matrix using Eq. (3.28) becomes $\Delta k_e = k_e$. Combining these two changes in the stiffness matrix, with

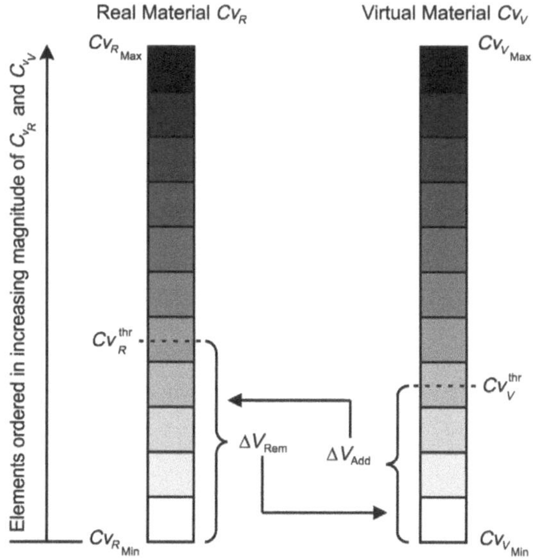

FIGURE 3.9 Ordered "*real*" and "*virtual*" material criterion domains for a minimum compliance design case, showing the criterion threshold and the volumes of material to be rearranged.

Eq. (3.27) give that the elemental criterion for the "*real*" and "*virtual*" material domains is then given by Eqs. (3.29) and (3.30), respectively.

$$Cv_{e_R} = \alpha_{e_R} = -u_e^T k_e u_e \tag{3.29}$$

$$Cv_{e_V} = \alpha_{e_V} = u_e^T k_e u_e \tag{3.30}$$

As the objective is to minimize the compliance of the design, the elements with the lower elemental criterion value are the ones to be added and removed from the "*real*" and "*virtual*" design domains. The real and virtual elements are ordered in increasing elemental criterion value, as shown in Fig. 3.9. Using the volume fractions to be added (ΔV_{Add}) and removed (ΔV_{Rem}), the values for the "*real*" and "*virtual*" material domains threshold values (Cv_R^{thr} and Cv_V^{thr}) are then calculated. It is important to note that these elemental criterion (Cv_e) values for both the "*real*" and "*virtual*" material domains are always positive for structural problems, but can be both positive and negative for compliant mechanism problems.

3.8.3 Elemental Criterion for Multiple Criteria

The elemental criterion (Cv_e) required for optimization problems with multiple criteria, such as multiple loading or multiple active support conditions, can be generated by combining the different normalized elemental criterion using the weighting factor approach of Eq. (3.31).

$$Cv_e = \sum_{k=1}^{K} Cv^*_{e_k}\omega_k, \quad \text{where: } \sum_{k=1}^{K}\omega_k = 1 \tag{3.31}$$

where Cv_e is the new combined elemental criterion for the eth element, K is the total number of criteria, $Cv^*_{e_k}$ is the normalized elemental criterion for the eth element of the kth criteria given by Eq. (3.32), ω_k is the weighting factor for the kth criteria and $Cv_{k_{\text{Max}}}$ is the maximum value of the kth criteria.

$$Cv^*_{e_k} = \frac{Cv_{e_k}}{Cv_{k_{\text{Max}}}} \tag{3.32}$$

3.8.4 Elemental Criterion for Compliant Mechanisms

Similar to the criterion for the minimum compliance problem, there are two elemental criteria (Cv_e) required for the case of optimizing a single input−output compliant mechanism; these are given by Eqs. (3.33) and (3.34) [34].

$$Cv_{e_R} = \alpha_{e_R} = u^T_{e,1}k_e u_{e,2} \tag{3.33}$$

$$Cv_{e_V} = \alpha_{e_V} = -\, u^T_{e,1}k_e u_{e,2} \tag{3.34}$$

where $u_{e,1}$ is the displacement field in the eth element due to the input force F_1 applied at the input port, and $u_{e,2}$ is the displacement field in the eth element due to the unit output force F_2 applied as a separate load case at the output port.

As the objective is to maximize the MPE of the compliant mechanism, the elements with the higher elemental criterion value are the ones to be added and removed from the "*real*" and "*virtual*" design domains. The real and virtual elements are ordered in increasing elemental criterion value, as shown in Fig. 3.10. Using the volume fractions to be added (ΔV_{Add}) and removed (ΔV_{Rem}), the values for the "*real*" and "*virtual*" material domains threshold values (Cv^{thr}_R and Cv^{thr}_V) are then calculated.

For the case of a compliant mechanism with multiple inputs and outputs, the multiple criteria Eq. (3.31) is then used to generate the two elemental criteria ($Cv_{e_{R\,\text{or}\,V}}$) of Eqs. (3.35) and (3.36).

$$Cv_{e_R} = \alpha_{e_R} = \sum_{i=1}^{n}\sum_{j=1}^{m}\omega_{i,j}(u^T_{e,i}k_e u_{e,j}) \tag{3.35}$$

$$Cv_{e_V} = \alpha_{e_V} = -\sum_{i=1}^{n}\sum_{j=1}^{m}\omega_{i,j}(u^T_{e,1,i}k_e u_{e,2,j}) \tag{3.36}$$

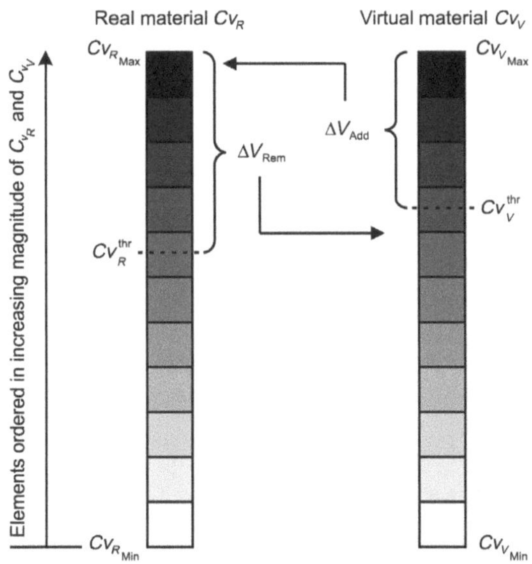

FIGURE 3.10 Ordered "*real*" and "*virtual*" material criterion domains for a compliant mechanism design case, showing the criterion threshold and the volumes of material to be rearranged.

where i corresponds to the ith input port P_i; j corresponds to the jth output port P_j; $u_{e,i}$ is the displacement field of the eth element due to the input force F_i, and $u_{e,j}$ is the displacement fields of the eth element due to the unit output force F_j.

As for the case of single input–output compliant mechanism, the objective is to maximize the MPE. Therefore, the same process of adding and removing elements shown in Fig. 3.10 is required for MIMO, i.e., the elements with the higher elemental criterion value are the ones to be added and removed from the real and virtual design domains.

3.9 MESH INDEPENDENT FILTERING

The mesh independent filter used Eq. (3.37) is based on the technique proposed by Sigmund and Petersson [37]. The criterion value of the eth element (Cv_e) is modified using Eq. (3.37) to become (Cv'_e), by calculating the weighted average of the criterion of only those elements which lie within a distance $r_{\min}$ (Eq. 3.38) of the eth element.

$$Cv'_e = \frac{\sum_{i=1}^{o} \mu_i \rho_i Cv_i}{\sum_{i=1}^{o} \mu_i} \tag{3.37}$$

$$\begin{aligned} &\mu_i = r_{\min} - \text{dist}(e, i) \\ &\{i \in O | \text{dist}(e, i) \leq r_{\min}\} \\ &i = 1, \ldots, O \\ &e = 1, \ldots, N \end{aligned} \tag{3.38}$$

where Cv'_e is the eth element filtered criterion; N is the number of finite elements in the mesh; ρ_i is the ith element virtual density; Cv_i is the ith element criterion; $r_{\min}$ is the length of the filter radius specified by the user, which needs to be larger than the smallest FE size in order be effective; dist(e, i) is the distance between the centers of the eth and ith elements; μ_i is the weighting factor for the ith element, its value decreases linearly the further the ith element is away from the eth element and has a value of zero for all elements outside of the filter radius, and O is the number of FE inside of the filter radius.

3.10 CONVERGENCE CRITERION

The convergence criterion is defined as the change in the objective function in the last 10 iterations and is given by Eq. (3.39). This number of iterations was found to be an adequate number of iterations for convergence to take place. It implies that the process will have a minimum of 10 iterations as the convergence criterion is not applied until the 10th iteration.

$$\varepsilon_i = \frac{\left| \sum_{i-9}^{i-5} OF_i - \sum_{i-4}^{i} OF_i \right|}{\sum_{i-4}^{i} OF_i} \tag{3.39}$$

where i is the current iteration number (> 10); OF_i is the objective function value in the ith iteration; ε_i is the convergence value of the objective function in the ith iteration.

REFERENCES

[1] M.P. Bendsøe, N. Kikuchi, Generating optimal topologies in structural design using a homogenization method, Comp. Meth. Applied. Mech. Eng. 71 (1988) 197–224.

[2] M.P. Bendsøe, Optimal shape design as a material distribution problem, Struct. Optim. 1 (1989) 193–202.

[3] M. Zhou, G. Rozvany, The COC algorithm, part II: topological, geometrical and generalized shape optimization, Comp. Meth. Appl. Mech. Eng. 89 (1991) 309–336.

[4] M.P. Bendsøe, O. Sigmund, Material interpolation schemes in topology optimization, Arch. Appl. Mech. 69 (1999) 635–654.

[5] H.A. Eschenauer, N. Olhoff, Topology optimization of continuum structures: a review, Appl. Mech. Rev. 54 (4) (2001) 331–390. Available from: http://dx.doi.org/10.1115/1.1388075.

[6] G.I.N. Rozvany (Ed.), Optimization of large structural systems, Nato – ASI, Lecture Notes, vol. 231, doi: 10.1007/978-94-010-9577-8, 1991.

[7] G.I.N. Rozvany, M.P. Bendsøe, U. Kirsch, Layout optimization of structures, Appl. Mech. Rev. 48 (2) (1995) 41–119.
[8] M.P. Bendsøe, O. Sigmund, Topology optimization theory, Methods and Applications, Springer, Berlin, 2003, ISBN: 3-540-42992-1.
[9] G.I.N. Rozvany, T. Lewinski (Eds.), Topology Optimization in Structural and Continuum Mechanics, Springer-VS, (CISM International Centre for Mechanical Sciences, vol. 549), ISBN 978-3-7091-1642-5, 2014.
[10] Y.M. Xie, G.P. Steven, A simple evolutionary procedure for structural optimization, Comput. Struct. 49 (1993) 885–896.
[11] O.M. Querin, G.P. Steven, Y.M. Xie, Evolutionary structural optimisation using an additive algorithm, Finite Elem. Anal. Des. 34 (2000) 291–308.
[12] O.M. Querin, Evolutionary structural optimisation stress based formulation and implementation, PhD dissertation, University of Sydney, 1997.
[13] O.M. Querin, G.P. Steven, Y.M. Xie, Evolutionary structural optimisation (ESO) using a bidirectional algorithm, Eng. Comput. 15 (8) (1998) 1031–1048.
[14] D. Reynolds, J. McConnachie, P. Bettess, W.C. Christie, J.W. Bull, Reverse adaptivity – a new evolutionary tool for structural optimization, Int. J. Numer. Meth. Eng. 45 (1999) 529–552.
[15] J.S. Liu, G.T. Parks, P.J. Clarkson, Metamorphic development: a new topology optimization method for continuum structures, Struct. Multidisc. Optim. 20 (4) (2000) 288–300.
[16] C.D. Chapman, K. Saitou, M.J. Jakiela, Genetic algorithms as an approach to configuration and topology designs, Trans. ASME J. Mech. Des 116 (1994) 1005–1013.
[17] S. Osher, J.A. Sethian, Fronts propagating with curvature-dependent speed: algorithms based on Hamilton–Jacobi formulations, J. Comput. Phys. 79 (1988) 12–49.
[18] J.A. Sethian, Level set methods and fast marching methods: evolving interfaces in computational geometry, fluid mechanics, Computer Vision, and Materials Science, Cambridge University Press, 1999, ISBN 0-521-64557-3.
[19] J. Sethian, A. Wiegmann, Structural boundary design via level set and immersed interface methods, J. Comput. Phys. 163 (2000) 489–528.
[20] M. Yulin, W. Xiaoming, A level set method for structural topology optimization and its applications, Adv. Eng. Softw. 35 (2004) 415–441.
[21] Y.M. Xie, G.P. Steven, Evolutionary structural optimization, Springer, London, 1997.
[22] G.I.N. Rozvany, O.M. Querin, Combining "ESO" with rigorous optimality criteria, Proceedings of OptiCON2000, Altair Engineering, Newport Beach, CA, USA, Oct. 26–27, Paper 6.
[23] G.I.N. Rozvany, O.M. Querin, Combining ESO with rigorous optimality criteria, Int. J. Vehicle Des. 28 (4) (2002) 294–299.
[24] G.I.N. Rozvany, O.M. Querin, Theoretical foundations of sequential Element Rejections and Admissions (SERA) methods and their computational implementations in topology optimisation, in: Proc. 9th AIAA/ISSMO Symposium on Multidisciplinary Analysis and Optimisation. AIAA, Reston, VA, doi:10.2514/6.2002-5521, 2002.
[25] G.I.N. Rozvany, A critical review of established methods of structural topology optimization, Struct. Multidisc. Optim. 37 (3) (2009) 217–237.
[26] G.I.N. Rozvany, O.M. Querin, Present limitations and possible improvements of SERA (Sequential Element Rejections and Admissions) methods in topology optimization, in: The Fourth World Congress of Structural and Multidisciplinary Optimization (WCSMO-4), Dalian, China, June 4–8, 2001.
[27] M. Zhou, G.I.N. Rozvany, On the validity of ESO type methods in topology optimisation, Struct. Multidisc. Optim. 21 (2001) 80–83.

[28] R.N. Brodie, O.M. Querin, D.C. Barton, Topology optimisation using the SERA method with a BESO bias on element admission, in: Proceedings of the 10th International Conference on Civil, Structural and Environmental Engineering Computing, Civil-Comp 2005.

[29] R.N. Brodie, O.M. Querin, D.C. Barton, G.I.N. Rozvany, Development and application of the SERA method for topology optimisation, in: 5th ASMO UK / ISSMO conference on Engineering Design Optimization, Stratford-upon-Avon, July 12–13, 2004, Paper 9.

[30] G.I.N. Rozvany, O.M. Querin, J. Logo, Sequential Element Rejection and Admission (SERA) method: application to multiconstraint problems, in: 10th AIAA/ISSMO Multidisciplinary Analysis and Optimization (MAO), Aug. 30–Sep. 01, 2004, Albany, New York, Paper AIAA-2004-4523, doi: 10.2514/6.2004-4523.

[31] R.N. Brodie, Development of controllability and robustness methodologies for Bi-Directional Evolutionary Structural Optimisation (BESO). PhD thesis, School of Mechanical Engineering, University of Leeds, UK, 2007.

[32] C. Alonso, R. Ansola, O.M. Querin, J. Canales, Design of compliant mechanisms with a Sequential Element Rejection and Admission method, in: Proceedings of the 10th World Congress on Computational Mechanics (WCCM2012), ISBN: 978-85-86686-70-2, Paper Nr. 19501, Sao Paulo, Brazil, 2012.

[33] C. Alonso Gordoa, R. Ansola, O.M. Querin, E. Vegueria, Parameter study of a SERA method to design compliant mechanism, in: Proceedings of the 14th AIAA/ISSMO multi-disciplinary analysis and optimization conference, September 17th–19th, 2012. Indianapolis, US, eISBN: 978-1-60086-930-3, doi:10.2514/6.2012-5525.

[34] C. Alonso, O.M. Querin, R. Ansola, A sequential element rejection and admission (SERA) method for compliant mechanisms design, Struct. Multidisc. Optim. 47 (6) (2013) 795–807.

[35] C. Alonso, Topology optimization of multidisciplinary compliant mechanisms with a Sequential Element Rejection and Admission method. PhD Dissertation, Universidad del País Vasco/Euskal Herriko Unibertsitatea/University of the Basque Country, Spain, 2013.

[36] R. Shield, W. Prager, Optimal structural design for given deflection, J. Appl. Math. Phys. 21 (1970) 513–523.

[37] O. Sigmund, J. Petersson, Numerical instabilities in topology optimization: a survey on procedures dealing with checkerboards, mesh-dependencies and local minima, Struct. Optim. 16 (1998) 68–75.

FURTHER READING

E. Hinton, J. Sienz, Fully stressed topological design of structures using an evolutionary approach, Eng. Comput., 12, 1995, pp. 229–244.

C. Mattheck, Design in Nature. Learning From Trees, Springer-Verlag Berlin Heidelberg, 1998.

G.I.N. Rozvany, M. Zhou, Applications of the coc algorithm in layout optimization, in: H. Eschenauer, C. Mattech, N. Olhoff (Eds.), Engineering Optimization in Design Processes, Proceedings International Conference Held in Karlsruhe, Germany, 1990, Springer, Berlin, 1991, pp. 59–70.

A. Baumgartner, L. Harzheim, C. Mattheck, SKO (soft kill option): the biological way to find an optimum structure topology, Int. J. Fatigue 14 (6) (1992) 387–393.

Chapter 4

Continuous Method of Structural Optimization

4.1 INTRODUCTION

Lin and Chao [1] considered topology optimization to be an extension of shape optimization. Where firstly, the topology of the structure is optimized followed by shape optimization. In another study [2], the opposite was done, where first the shape was optimized followed by topology optimization using the Homogenization method [3] to determine the material orientation. Irrespective of the order in which the optimization was carried out; shape first followed by topology, or vice versa; the shape is dependent on the material distribution, and equally the material distribution is dependent on the shape. For this reason, an iterative approach of shape followed by topology optimization; or vice versa; can be demonstrated to be more efficient than just using one method, and only when that one finishes, then using the other [2], especially if small changes in the shape or topology can have a significant effect on the behaviour of the structure. Due to the interaction between structural behaviour and its shape/topology, the use of isolines to show the behaviour and control the shape/topology can allow for a faster convergence toward an improved design.

The relationship between the desired field contours represented by the isolines, and the shape/topology of the design is suggested in several studies [4–7]. In Hassani and Hinton [4], the contours of the material density obtained from topology optimization are approximated by Bezier curves [8]. Through a threshold method, the obtained contour is adopted as the boundary of the structure, which can be used either as a new design for the next iteration or as an initial design for shape optimization. In the study of Woon et al. [9], a relationship between the field contours in the form of stress isolines and the shape and location of cavities was shown. The use of isolines was also used by Kim et al. [10,11] for the implementation of fixed grid finite element analysis (FEA) into the Evolutionary Structural Optimization (ESO) method and by Cui et al. [12] in the extended ESO (XESO) method to allow removal and admission of material.

Topology Design Methods for Structural Optimization.
DOI: http://dx.doi.org/10.1016/B978-0-08-100916-1.00004-0

The procedure for material removal/addition plays a crucial role in finding the optimal design. The Isolines Topology Design (ITD) method was developed as an integrated procedure that is capable of designing both the shape and topology of two and three-dimensional (2D & 3D) continuum structures by using the isolines or isosurfaces of the desired structural behaviour of that structure. This chapter presents the work of Victoria et al. [13–19] on the implementation of topology design using the isolines and isosurfaces method.

4.2 THE ISOLINES TOPOLOGY DESIGN ALGORITHM

The ITD algorithm works in an iterative manner, by redistributing material (adding and removing) inside of the design domain, until it reaches a desired volume fraction. The ITD algorithm for 2D and 3D structures consists of the following 11 steps. A schematic representation of this algorithm is given in Fig. 4.1

1. Define the design problem: This consists of both the design and nondesign domains; the structural supports and loads; the material properties; the finite element (FE) mesh; the optimization problem; and the design criterion (C_{Des}), (Section 4.3).
2. Specify the ITD parameters: These consist of the target final design volume (V_{F}); the total number of iterations (N_{Tot}); the total number of load cases (N_{LC}); the total number of material phases (N_{MP}); the volume control weighting factor for the different material phases (β^{MP}) and the minimum volume change limit ($\Delta V_{\text{Min}_{\text{Lim}}}$) or ($\% V_{\text{Min}_{\text{Lim}}}$) (Section 4.4).
3. Carry out the initial analysis of the design domain using the FE Method (FEM) for the total number of load cases (N_{LC}), (section 4.5).
4. Determine the target volume (V_i) for the current (ith) iteration, (section 4.6).
5. Calculate the Minimum Criterion Level (MCL) for the current (ith) iteration, (section 4.7).
6. Reset the stabilization process iteration number ($k = 1$)
7. Extract the shape (2D) or surface (3D) for the design (Section 4.8).
8. If the minimum volume change limit ($\Delta V_{\text{Min}_{\text{Lim}}}$) or ($\% V_{\text{Min}_{\text{Lim}}}$) is reached, go to step 10, otherwise go to step 9 (Section 4.9).
9. Carry out the reanalysis of the design domain using the FEM (Section 4.5), increment the stabilization process iteration number ($k = k + 1$) and go to step 7.
10. If the total number of iterations (N_{Tot}) is reached, go to step 11, otherwise, increment the iteration number ($i = i + 1$) and go to step 4.
11. Stop the design process.

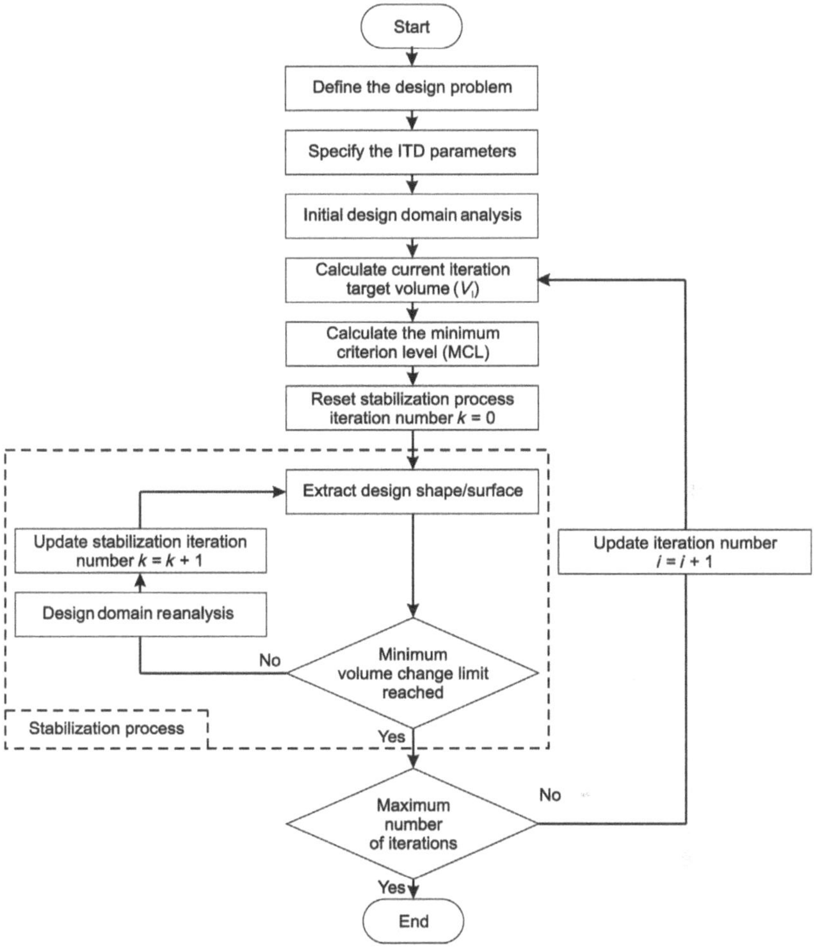

FIGURE 4.1 Flow chart of the general Isolines/Isosurfaces Topology Design (ITD) algorithm.

4.3 THE OPTIMIZATION PROBLEM

The optimization problem consists of minimizing the selected design criterion (Eq. 4.1) within the design domain, subject to achieving a target volume (Eq. 4.2), structural equilibrium (Eq. 4.3) and the ITD inequality condition which must be satisfied by all the nodes of the structure (Eq. 4.4).

$$\text{minimize} \quad (C_{\text{Des}}) \tag{4.1}$$

$$\text{subject to} \quad V = \sum_{e=1}^{N} \xi^e V_e \leq V_{\text{F}}; \;\; e = 1, \ldots, N; \;\; 0 \leq \xi^e \leq 1 \tag{4.2}$$

$$[\mathbf{K}]\{\mathbf{u}\} - \{\mathbf{P}\} = 0 \tag{4.3}$$

$$Cv_{\text{MCL}} - NCv_n \leq 0 \tag{4.4}$$

where (C_{Des}) is the selected design criterion; (V) is the volume of the structure within the design domain; (ξ^e) is the fraction of the eth FE in the structure which is inside of the structural domain (Section 4.5); (V_e) is the volume of the eth FE in the structure; (N) is the total number of FE in the structure; (**K**) is the stiffness matrix for the structure; (**u**) is the displacement vector of the structure; (**P**) is the applied nodal load vector; (Cv_{MCL}) is the value of the minimum criterion level, (section 4.7); (n) is the nth node of the FE mesh which lies inside or on the boundary of the structure, and (NCv_n) is the value of the design criterion in the nth node of the structure.

4.3.1 Criterion Selection

The design criterion (C_{Des}) used throughout this work was the von Mises stress (σ_{vM}), which for a general continuum domain is calculated using (4.5).

$$\sigma_{vM} = \sqrt{\frac{1}{2}\left[(\sigma_1-\sigma_2)^2 + (\sigma_2-\sigma_3)^2 + (\sigma_1-\sigma_3)^2\right]} \tag{4.5}$$

where σ_1, σ_2, and σ_3 are the principal stresses.

4.3.2 Criterion for Problems with Different Tensile and Compressive Structural Behaviour

A loaded structure will have regions where the material experiences: (1) a tensile field; (2) a compressive field; and (3) where both tensile and compressive fields may be present but acting orthogonally from each other in a biaxial or triaxial state of stresses. In general topology optimization, an underlying assumption is that the structural/material behaviour is the same in tensile and in compressive regions. However, this is not necessarily the case as two different conditions are possible:

1. The same material may behave differently in tension and compression; i.e., different Young's moduli or different yield stresses;
2. Different materials may be required for the tensile and compressive components of the structure; i.e., steel-reinforced concrete where the steel supports the tensile loads and the concrete the compressive loads;

In the literature [18], several approaches have been used to solve this type of problem, these include: optimality criteria; ground structure; growth method; SIMP; homogenization; and heuristics. The approach implemented in ITD is that of Querin et al. [18], which combines the works of Pálfi [20], Dewhurst [21], and Nair [22].

Pálfi [20] aligned the material orientation with the principal stress directions in each FE of the structure. Dewhurst [21] demonstrated that an optimal structural layout required constant strain energy per unit weight in all its members, where these members had to have a constant strain ratio as a function of their elastic moduli and densities. Nair [22] and Taggart et al. [23] used a node based approach to carry out the optimization using the strain energy density as the optimality criteria, assumed isotropic material properties for each element, but generated a fictitious isotropic elastic modulus for the biaxial stressed elements.

From Dewhurst [21], an optimal layout of maximum stiffness and minimum volume/weight is obtained if all the structural members have the constant fixed strain ratio of Eq. (4.6).

$$k = \frac{\varepsilon_T^*}{\varepsilon_C^*} = \left(\frac{\rho_T E_C}{\rho_C E_T}\right)^{\frac{1}{2}} \tag{4.6}$$

where k is the constant fix ratio between strains in the optimum structure; $\varepsilon_T^*, \varepsilon_C^*$ are the tensile and compressive maximum strains that define the optimal structure layouts; E_T, E_C are the elastic moduli of the tensile and compressive members of the structure; ρ_T, ρ_C are the density, mass, weight, or cost per unit volume of the tension and compressive members of the structure

If it is more convenient to work with stresses rather than strains, Eq. (4.6) can be rearranged by dividing with the respective modulus of elasticity, to give the equivalent equation Eq. (4.7), that relates the tensile and compressive stresses in the optimal structure by the new factor (k_1):

$$k_1 = \frac{\sigma_T^*}{\sigma_C^*} = \left(\frac{\rho_T E_T}{\rho_C E_C}\right)^{\frac{1}{2}} \tag{4.7}$$

where k_1 is the constant fix ratio between stresses in the optimum structure, and σ_T^*, σ_C^* are the tensile and compressive maximum stresses in the optimal structure.

In order to obtain an optimum using the optimality criteria of Dewhurst [21], a fully stressed design solution of the modified strain or stress field in Eq. (4.6) or (4.7) is required. For the work presented in this book, the stress field was used using the von Mises stress of Eq. (4.5), after the material properties in the structure had been modified using Eq. (4.6) or (4.7) and the negative principal stresses were multiplied by k_1, Section (4.5.4).

4.3.3 Nondesign Domain Region

A structural domain may contain regions which should not to be optimized. A possible reason for this may be due to how such a structure interacts with other structures or external loading requirements. Such regions which are not allowed to take part in the optimization are called nondesign regions. In the ITD algorithm, the material properties of the fixed grid for those elements in these region are not altered during the design process.

4.4 THE ITD PARAMETERS

There are six parameters which drive the ITD method, these are: (1) target final design volume (V_F); (2) total number of iterations (N_{Tot}); (3) total number of load cases (N_{LC}); (4) total number of material phases (N_{MP}); (5) volume control weighting factor for the different material phases (β^{MP}); and (6) minimum volume change limit ($\Delta V_{Min_{Lim}}$) or ($\% V_{Min_{Lim}}$).

4.4.1 Target Final Design Volume

The target final design volume (V_F), is the user specified final volume which the structure must have, when the optimization process has finished.

4.4.2 Total Number of Iterations

The total number of iterations (N_{Tot}) is also specified by the user. This value is used to calculate the target design volume (V_i) at each iteration of the optimization process, (section 4.6).

4.4.3 Total Number of Load Cases

Structures which are used in real environments are subjected to loads which may be applied simultaneously or at different times in the life of the structure. These types of loadings are referred to as multiple load cases and all must be considered in the design of a structure. When only one load case is applied ($N_{LC} = 1$), the standard ITD algorithm of Section 4.2 is applied to the optimization problem. However, for any other number of load cases ($N_{LC} > 1$), the MCL for multiple load cases of Section 4.7.2 needs to be used.

4.4.4 Total Number of Material Phases

A heterogeneous or multimaterial structure is referred to as a solid structure made of different materials distributed continuously or discontinuously. Its properties can be adjusted by controlling its material composition, microstructure, and geometry [24]. Most structures can be considered to be made from a single homogeneous material. However, there are instances when the application of a structure should have certain characteristics, such as heat resistance, antioxidation properties, high-yield stress or ultimate tensile strength. Such properties may not be obtained by the use of a single or homogeneous material. Such structures, with varying properties across the spatial domain, are commonly used in aerospace; biomedical; civil; geophysical; and nanoscale structures. In these applications, their performance objectives are achieved by the capability of the varying material properties acting both globally and locally across the design domain. When only one material phase is required ($N_{MP} = 1$), the standard ITD algorithm of Section 4.2 is applied to the optimization problem. However, for any other number of material phases ($N_{MP} > 1$), the MCL for multiple material phases defined in Section 4.7.3 needs to be used.

4.4.5 The Weighting Factor for the Different Material Phases

When the structure to be optimized consists of multiple material phases ($N_{MP} > 1$), the volume fraction (β^{MP}) for each of the phases needs to be specified by the designer. This provides the value of the required volume fraction of each material phase in the final design domain. As it is a volume fraction, it must satisfy Eq. (4.8).

$$\sum_{j=1}^{N_{MP}} \beta_j^{MP} = 1 \tag{4.8}$$

where j corresponds to the jth material phase used in the design domain.

4.4.6 The Minimum Volume Change Limit

Any change to the structural boundary due to the optimization process, will have an effects on the criterion distribution. The minimum volume change limit parameter ($\Delta V_{\text{Min}_{\text{Lim}}}$) or ($\% V_{\text{Min}_{\text{Lim}}}$), controls when the optimization process can continue. The typical value used throughout the examples in this book is ($\Delta V_{\text{Min}_{\text{Lim}}} = 0.01$) or ($\% V_{\text{Min}_{\text{Lim}}} = 1\%$).

4.5 ANALYSIS OF THE DESIGN DOMAIN

The analysis of the design domain consists of determining the behaviour of the structure due to the applied loading, and this is most commonly carried out using the FE method. In the ITD algorithm, the boundary of the structure is manipulated in order to satisfy the ITD inequality constraint. This differs from other topology optimization methods, where the FEs themselves are the design variables. However, the cost of implementing a new FE mesh every time the boundary changed is too costly. Instead, the Fixed Grid Finite Element Method (FG-FEM) was applied. The ITD algorithm requires an initial analysis of the design domain for the optimization process to start, followed by repetitive reanalysis at every iteration. For most problem applications, these two analyses are exactly the same, except for problems with different tensile and compressive structural behaviour.

4.5.1 Fixed Grid Finite Element Method

The FG-FEM was first introduced by García-Ruiz and Steven [25] as a tool for numerical estimation of two-dimensional (2D) elasticity problems, and was later extended to three-dimensional (3D) structures by Garcia et al. [26,27]. The benefits of using the FG-Finite Element Analysis (FG-FEA) over conventional FEA in ITD are that: (1) FG does not need a fitted mesh to discretize the design domain; (2) the boundary of the design is disassociated from the mesh [25]; (3) designs using FG-FEA do not contain checkerboard patterns, making the design more reliable for manufacture [28]; and (4) solution time is significantly reduced [29]. In FG-FEA, the elements are in a fixed position and have the real design superimposed on them (Fig. 4.2). This means that there are elements which lie inside (I), outside (O), or on the boundary (B) of the design [10,25].

Due to the fixed grid geometry of all the finite elements, the stiffness matrix for each element is almost fixed and only depends on its material properties (elasticity module). For the case of inside and outside elements, these properties are constant. For boundary elements, the properties consist of a combination of the inside (I) and outside (O) materials. The fixed grid approximation then transforms the bimaterial element into a homogeneous

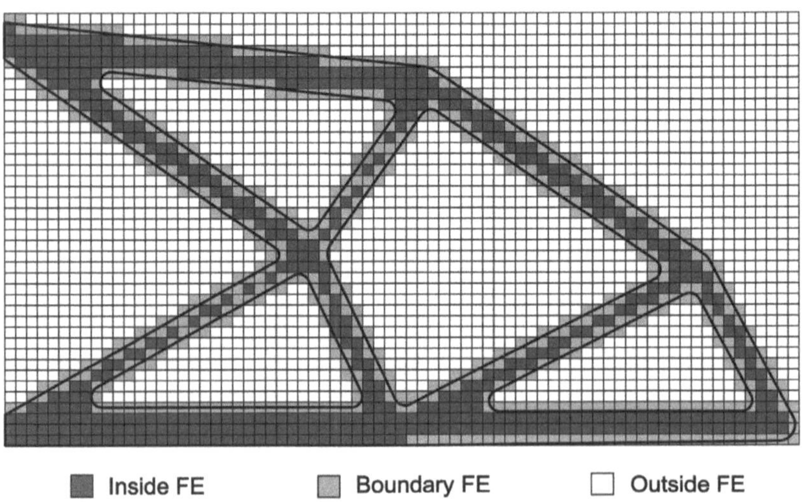

FIGURE 4.2 Fixed grid approximation of a structure, with the finite elements classified according to their position in relations with the boundary between the isolines and the structure.

isotropic element where the material property is scaled by function (4.9), which is the ratio of volume of inside (I) material to the total material within that element.

$$\xi^e = \frac{V_{\mathrm{I}}^e}{V_{\mathrm{I}}^e + V_{\mathrm{O}}^e} = \frac{V_{\mathrm{I}}^e}{V^e} \tag{4.9}$$

where ξ^e is the design fraction of inside to total volume of the eth FE in the structure; V_{I}^e is the volume of material of the eth FE inside of the structural boundary; V_{O}^e is the volume of material of the eth FE outside of the structural boundary; and V^e is the total volume for the eth finite element.

The stiffness matrix for each finite element is always linearly proportional to ξ^e. This is the case when either a fixed grid domain of equal sized FE is used or an unstructured FE mesh is used. This greatly reduces the time taken to modify the stiffness matrix every time the boundary changes, as only those FE which have changed properties have their stiffness matrix modified. The elemental stiffness matrix K^e is given by Eq. (4.10).

$$K^e = \begin{cases} K_I & \text{if} \quad \xi^e = 1 \\ K_O & \text{if} \quad \xi^e = 0 \\ K_B = K_I\xi^e + K_O(1 - \xi^e) & \text{if} \quad 0 < \xi^e < 1 \end{cases} \tag{4.10}$$

where K^e is the elemental stiffness matrix of the eth FE in the structure; ξ^e is the design fraction of inside to total volume of the eth FE in the structure, given by Eq. (4.9), K_I is the stiffness matrix for an element inside of the structural boundary; K_O the stiffness matrix for an element outside of the structural boundary, normally $K_O = K_i \times 10^{-6}$, and K_B is the stiffness matrix for an element on the boundary of the structure.

4.5.2 Calculating the Elemental Criterion Value

For each load case, the criterion value for an element (Cv_e) is calculated as a weighted average of the design criterion values (C_{Des}) at the Gauss points of that element, given by Eq. (4.11).

$$Cv_e = \frac{\sum_{g=1}^{N_G} \omega_g^e C_{\mathrm{Des}_g}^e}{\sum_{g=1}^{N_G} \omega_g^e} \tag{4.11}$$

where N_G is the number of the Gauss points in the eth element; ω_g^e is the weighting factor for the gth Gauss point of the eth element; and $C_{\mathrm{Des}_g}^e$ is the design criterion value at the gth Gauss point of the eth finite element.

4.5.3 Calculating the Nodal Criterion Value

For each load case, the criterion value for the nth node in the structure (NCv_n) is calculated as the average of the nodal criterion values (NCv_i^e) of all the finite elements attached to the nth node, given by Eq. (4.12).

$$NCv_n = \frac{\sum_{e=1}^{N_{n\mathrm{Tot}}} NCv_n^e}{N_{n\mathrm{Tot}}} \tag{4.12}$$

where n is the nth node number of the FE mesh; NCv_n^e is the nodal criterion value in the eth FE attached to the nth node; and $N_{n\mathrm{Tot}}$ is the total number of FE connected to the nth node. The nodal criterion value for the eth FE (NCv_n^e) is determined from the criterion values at each Gauss extrapolated to the nth nodes using the shape functions of the element.

4.5.4 Initial Design Domain Analysis

For the ITD algorithm to commence, it requires an initial evaluation of the structural behaviour of the design domain to the applied loading. The nature of this analysis depends on the particular problem to be optimized, of which there are two general types: (1) with applied single or multiple load cases and with single or multiple material phases and (2) where the structure exhibits a different tensile and compressive behaviour.

4.5.4.1 For Problems With Multiple Load Cases and Material Phases

A standard finite element analysis is all that is required for the initial analysis of the design domain, Fig. 4.3, when a structural domain has applied either a single or multiple load cases, or when the material phase required is either single or multiple. For ITD, this corresponds with using the FG-FEM of Sections 4.5.1–4.5.3.

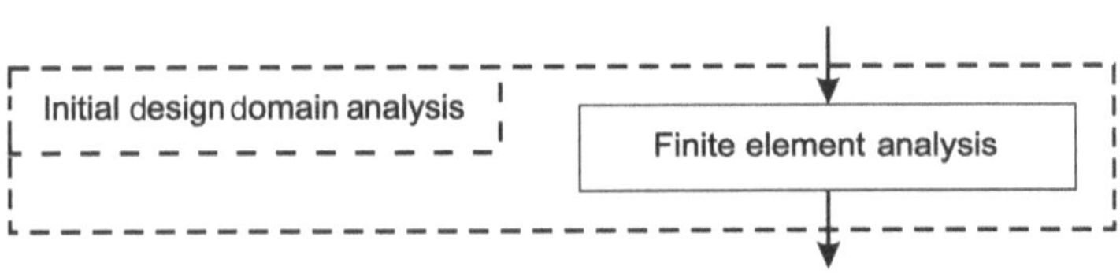

FIGURE 4.3 Schematic diagram showing the type of initial analysis required for problems with applied single or multiple load cases and/or with single or multiple material phases.

4.5.4.2 For Problems with Different Tensile and Compressive Structural Behaviour

The analysis required for problems where the structure exhibits a different behaviour in tensile and compressive regions is much more elaborate, and two cases need to be considered:

1. The use of different materials for the tensile and compressive regions (elements) of the structure (e.g., steel-reinforced concrete), or
2. The use of the same material for the tensile and compressive regions (elements) of the structure.

Case 1: Different Material for Tensile and Compressive Regions

Since the material properties (ρ and E) are known for the two materials, these are then used in the analysis of the structure. Once the strains or stresses are calculated in each element, all of the negative terms are then multiplied by the factor k (Eq. 4.6) or k_1 (Eq. 4.7), which were calculated using the known material properties. The resulting strain or stress field is then used to calculate the design criterion, which for the work presented in this book, was the von Mises stress (Eq. 4.5).

Case 2: The Same Material for Tensile and Compressive Regions

Since the same material is used throughout the structure, the ratio of the densities in Eqs. (4.6) and (4.7) can be set equal to unity. The factor k or k_1 is specified by the user, and the artificial modulus of elasticity of Eq. (4.13) is calculated for all of the compressive elements of the structure.

$$E_C = \frac{E_T}{k_1^2} = E_T k^2 \tag{4.13}$$

Once the modulus of elasticity have been updated for all compressive elements of the structure using Eq. (4.13), the structure is analyzed, and all negative strains or stresses are multiplied by the factor k or k_1, respectively. The resulting strain or stress field is then used to calculate the design criterion, which for the work presented in this book, was the von Mises stress (Eq. 4.5).

Since the structure has different real or artificially generated mechanical properties for its tensile and compressive elements, the analysis has to capture this. For a two-dimensional (2D) or three-dimensional (3D) continua, an orthotropic analysis needs to be carried out. However, when the ITD process is started, it is impossible to determine which components of the structure experiences tensile or compressive stresses or both. The following two stage process then needs to be implemented which consists of an:

1. Initial orthotropic material identification stage;
2. Orthotropic analysis and stress manipulation stage.

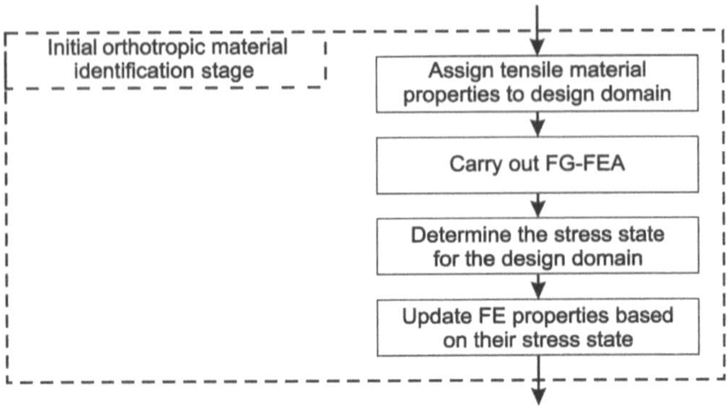

FIGURE 4.4 Schematic diagram showing the initial orthotropic material identification stage.

The initial orthotropic material identification stage, Fig. 4.4, consists of the following four steps:

1. Assign the tensile material property to the entire design domain;
2. Carry out an analysis of the design domain using the FG-FEA;
3. Determine the biaxial (2D) or triaxial (3D) stress state for the design domain by calculating the principal stresses (σ_1, σ_2, and σ_3 for 3D) and their orientations in each FE [30]. For the 3D continuum case, three stress states can exist: (a) all stresses are positive ($\sigma_1 > \sigma_2 > \sigma_3 > 0$); (b) all stresses are negative ($0 > \sigma_1 > \sigma_2 > \sigma_3$), and (c) one or more of the stresses are negative ($\sigma_1 > \sigma_2 > 0 > \sigma_3$) or ($\sigma_1 > 0 > \sigma_2 > \sigma_3$).
4. Update the material property of each FE in the design domain depending on the stress state satisfied. For all elements satisfying case: (a) the material properties are set as isotropic and have the values of the tensile material; (b) the material properties are set as isotropic and have the values of the compressive material, and (c) the material properties set as orthotropic with the directions of positive stress given the values of the tensile material and the directions of negative stress given the values of the compressive material.

The orthotropic analysis and stress manipulation stage, Fig. 4.5, consists of the following five steps:

1. Carry out a structural analysis with the new quasiorthotropic material properties;
2. Calculate the principal stresses and their orientation.
3. Multiply all of the negative principal stresses by the factor k_1(Eq. 4.7).
4. Calculate the von Mises stress distribution using the positive principal stresses and the modified negative stresses.
5. Determine the new orthotropic (or isotropic) properties for each element of the structure using both the principal stresses and their orientations calculated in step 2, and the stress state satisfied (a, b, or c).

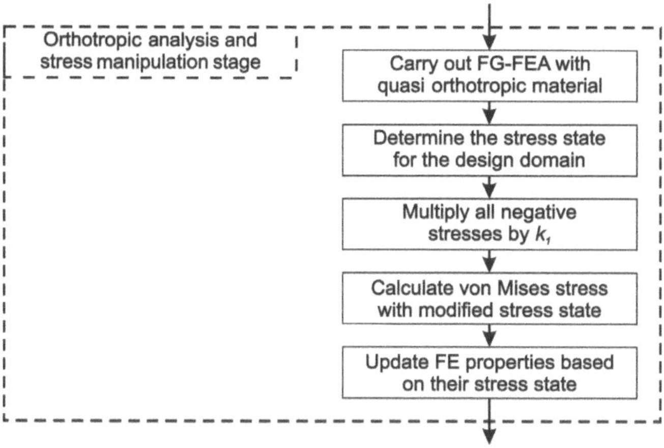

FIGURE 4.5 Schematic diagram showing the orthotropic analysis and stress manipulation stage.

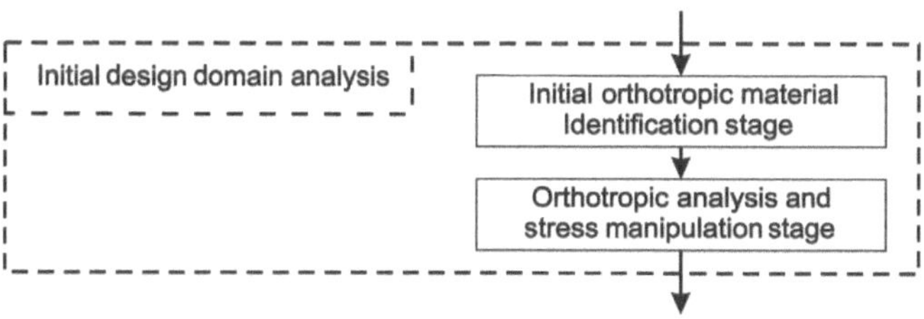

FIGURE 4.6 Schematic diagram showing the type of initial analysis required for problems with different tensile and compressive structural behaviour.

The initial analysis of the design domain for problems with different tensile and compressive structural behaviour consists of firstly carrying out the initial orthotropic material identification stage immediately followed by the orthotropic analysis and stress manipulation stage, Fig. 4.6.

4.5.5 Reanalysis of the Design Domain Analysis

During the stabilization process of the ITD algorithm, the structural behaviour of the design domain needs to be reevaluated. The nature of this analysis, in a similar way to the initial analysis, depends on the two types of problems to be optimized: (1) with applied single or multiple load cases and with single or multiple material phases and (2) where the structure exhibits a different tensile and compressive behaviour.

4.5.5.1 For Problems with Multiple Load Cases and Material Phases

A standard finite element analysis is all that is required for the reanalysis of the design domain, Fig. 4.7, when a structural domain has applied either a

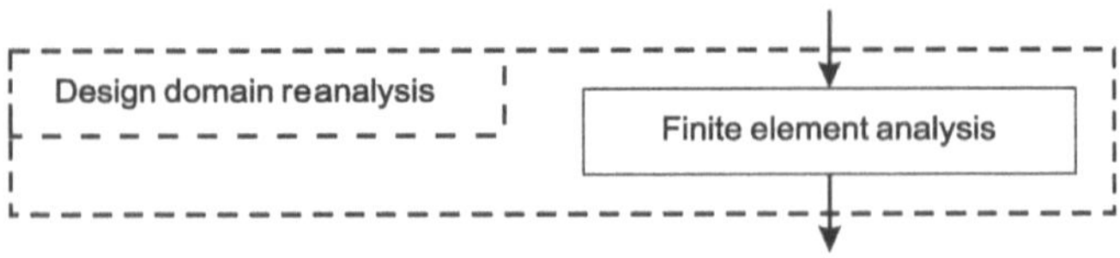

FIGURE 4.7 Schematic diagram showing the type of reanalysis required for problems with applied single or multiple load cases and/or with single or multiple material phases.

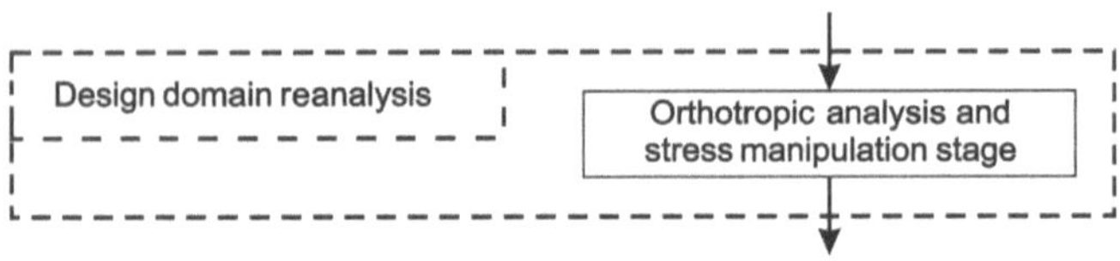

FIGURE 4.8 Schematic diagram showing the type of reanalysis required for problems with different tensile and compressive structural behaviour.

single or multiple load cases, or when the material phase required is either single or multiple. For ITD, this corresponds with using the FG-FEM of Sections 4.5.1–4.5.3.

4.5.5.2 For Problems With Different Tensile and Compressive Structural Behaviour

The reanalysis process for problems with different tensile and compressive structural behaviour, consists of only the 2nd stage of the initial analysis for the same type of problems. That is, the orthotropic analysis and stress manipulation stage, Fig. 4.8 explained in Section 4.5.4.2 and Fig. 4.5.

At the initial stages of the development of the ITD algorithm, two FEA were carried out during the orthotropic analysis and stress manipulation stage. The first to determine the sign and orientation of the principal stresses in order to calculate the orthotropic properties for each element. The second to determine the principal stresses to use for the optimization. However; apart from the initial analysis stage, it was found that changes in the structural domain due to the optimization step have a negligible effect in the changes of the direction and sign of the principal stresses. Hence only one finite element analysis is required in the reanalysis process of the orthotropic analysis and stress manipulation stage.

4.6 DETERMINING THE TARGET VOLUME

During each iteration of ITD, the target volume for that iteration is calculated using (Eq. 4.14).

$$V_i = V_0\left(\frac{N_{\mathrm{Tot}} - i}{N_{\mathrm{Tot}}}\right) + V_{\mathrm{F}}\left(\frac{i}{N_{\mathrm{Tot}}}\right) \tag{4.14}$$

where V_i is the target design volume for the ith iteration, (N_{Tot}) is the total number of iterations, V_0 is the initial volume, and V_{F} is the target final volume for the optimized structure set by the designer.

4.7 DETERMINING THE MINIMUM CRITERION LEVEL (MCL)

4.7.1 MCL for Single Load Case Problems

Once the criterion in the design domain is calculated for each element, these are arranged in decreasing order of criterion value, Fig. 4.9. An element by element volume summation of the ordered list is carried out until a volume is reached which is as close as possible to the target volume given by Eq. (4.14), where the error level between the summed and target volume depends on the size of the elements. The criterion value of the next element in the ordered list is then used as the value for the MCL,(Cv_{MCL}).

4.7.2 MCL for Multiple Load Case Problems

For ITD to deal with multiple load case problems, the process of superposition was used. This means that the designs obtained for each individual load case are superimposed "or added together" to produce the new design, whilst satisfying the target volume constraint. This technique is analogous to the AND/OR method applied to other methods such as ESO/BESO [7,31−33]. The algorithm for calculating the MCL for the multiple load case problems requires the following eleven steps, a schematic diagram of which, is given in Fig. 4.10.

1. For each load case calculate the average (or mean) criterion ($Cv_{\mathrm{lc}}^{\mathrm{mean}}$)
2. Calculate the target volume for each load case design domain ($V_{i_{\mathrm{lc}}}$) using (4.15).

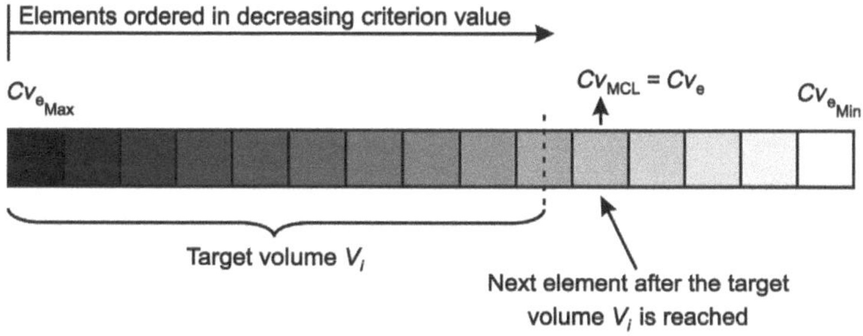

FIGURE 4.9 Schematic diagram showing how the MCL is determined as being the value of the selected criterion of the element after the target volume of elements is calculated.

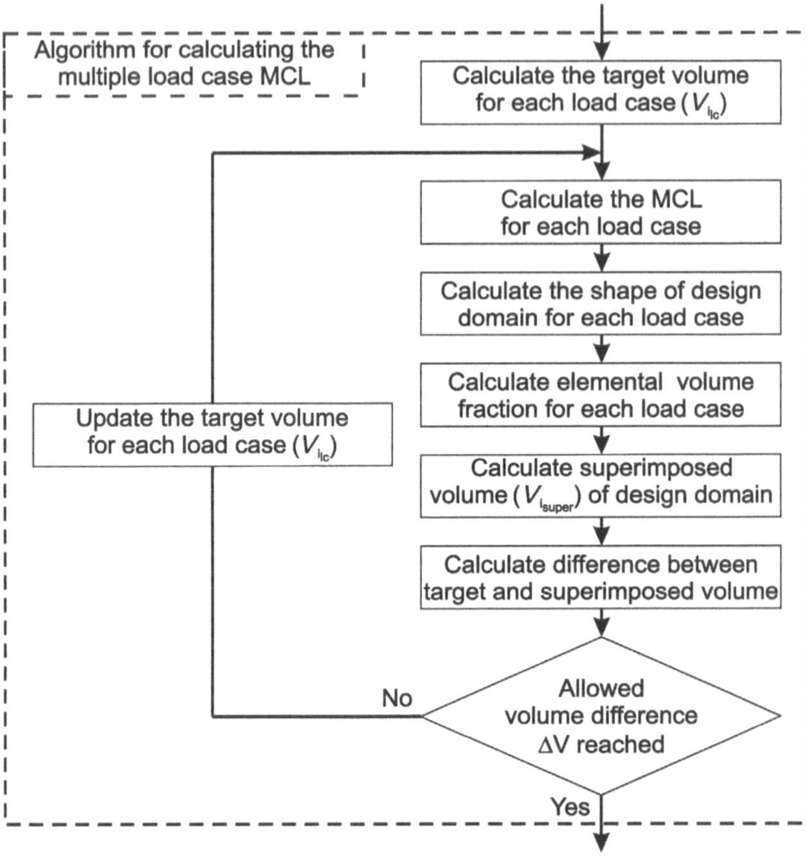

FIGURE 4.10 Schematic diagram showing the algorithm for calculating the MCL for multiple load case problems.

$$V_{i_{\mathrm{lc}}} = \frac{\beta_{\mathrm{lc}} \times V_i}{\sum_{\mathrm{lc}=1}^{N_{\mathrm{LC}}} \beta_{\mathrm{lc}}} \tag{4.15}$$

where i is the ith iteration; lc is the lc-th load case; N_{LC} is the total number of load cases; β_{lc} is the volume control weighting factor for the lc-th load case. It is calculated using the ratio of the average criterion for the lc-th load case with the maximum average criterion amongst all other load cases, $\max_{\mathrm{lc}=1,N_{\mathrm{LC}}}\{Cv_{\mathrm{lc}}^{\mathrm{mean}}\}$ and is given by (4.16).

$$\beta_{\mathrm{lc}} = \frac{Cv_{\mathrm{lc}}^{\mathrm{mean}}}{\max_{\mathrm{lc}=1,N_{\mathrm{LC}}}\{Cv_{\mathrm{lc}}^{\mathrm{mean}}\}} \tag{4.16}$$

3. Use the individual load case design target volume from Eq. (4.15) to calculate the MCL for each load case ($Cv_{\mathrm{MCL_{lc}}}$).
4. For each load case, calculate shape of the design domain using that load case MCL,($Cv_{\mathrm{MCL_{lc}}}$).
5. Use the shape of the produced design domain to calculate the volume fraction of each finite element for all of the load cases (ξ^e_{lc}).
6. The volume fraction for an element (ξ^e) is then given by the maximum volume for that element from all of the load cases Eq. (4.17).

$$\xi^e = \max_{\mathrm{lc}\,=\,1,N_{\mathrm{LC}}} \{\xi^e_{\mathrm{lc}}\} \tag{4.17}$$

7. Calculate the superimposed volume of the design domain using Eq. (4.18).

$$V_{i_{\mathrm{super}}} = \sum_{e=1}^{N} V_e \times \xi^e \tag{4.18}$$

where e is the eth finite element number; V_e is the volume of the eth element;

8. Calculate the volume difference between the target volume (V_i) and the superimposed volume ($V_{i_{\mathrm{super}}}$) using (4.19).

$$\Delta V = V_{i_{\mathrm{super}}} - V_i \tag{4.19}$$

9. If the absolute value of the volume error is less than the volume of an element ($|\Delta V| \leq V_e$), go to step 11 otherwise go to step 10. *Note that since this process was implemented using a fixed grid, the assumption is that all elements have the same volume* (V_e). *If the element size is not constant, then either the minimum* ($V_{e_{\min}}$) *or the average* ($V_{e_{\mathrm{avg}}}$) *element volume can be used to determine this limit.*
10. The volume required for each load case design domain ($V_{i_{\mathrm{lc}}}$) needs to be updated to that given by Eq. (4.20), go to step 3.

$$V_{i_{\mathrm{lc}}} = V_{i_{\mathrm{lc}}} - \beta_{\mathrm{lc}} \times \mathrm{sgn}(\Delta V) \times V_e \tag{4.20}$$

11. Continue with the optimization process

4.7.3 MCL for Multiple Material Phases Problems

For ITD to deal with multiple material phases, the process is straight forward. In order to produce a stiff structure, the regions of the structural domain with the high values of the design criterion need to be supported by the material phases with the highest stiffness and vice versa. The algorithm for calculating the MCL for the multiple material phases requires the following four steps, a schematic diagram of which, is given in Fig. 4.12.

1. Arrange the j-material phases ($j = 1, \ldots, N_{\mathrm{MP}}$) in decreasing order of stiffness by using their Young's moduli. The greatest value of the Young's modulus is indexed E_1 and the smallest value is indexed $E_{N_{\mathrm{MP}}}$.
2. Calculate the volume of every material phase (V_j) in the current ith iteration, using Eq. (4.21)

$$V_j = \beta_j^{\mathrm{MP}} \times V_i$$
$$\sum_{j=1}^{N_{\mathrm{MP}}} \beta_j^{\mathrm{MP}} = 1 \tag{4.21}$$

   where β_j^{MP} is the volume control weighting factor. Its value is initially proposed by the designer.
3. Calculate the MCL for each material phase Cv_{MCL_j} associated with V_{i_j} using (4.22), see Fig. 4.11.

$$V_{i_j} = \sum_{j=1}^{N_{\mathrm{MP}}} V_j \tag{4.22}$$

4. Determine the material properties for every FE (E^e), Fig. 4.11, using the design fraction inside the element (ξ^e), the minimum criterion level for each phase ($Cv_{\mathrm{MCL},j}$) and Eq. (4.23). Although in this analysis, only the modulus of elasticity was changed, all properties should be modified using the form of Eq. (4.23), these could consist of the material density of the element (ρ^e) or other physical properties required in the analysis.

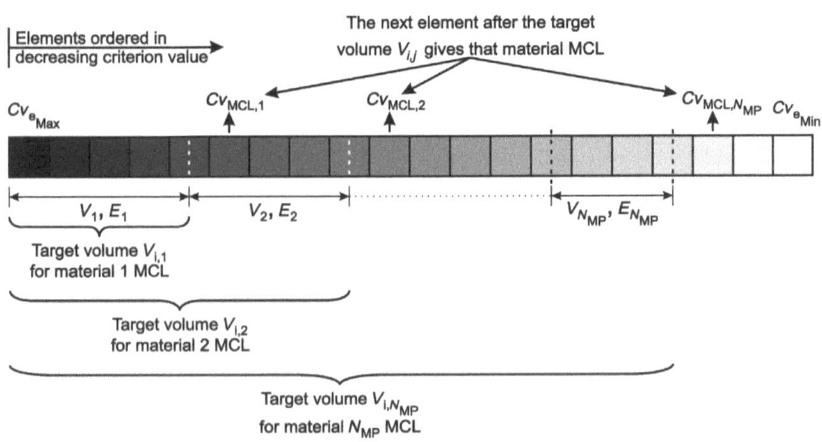

FIGURE 4.11 Schematic diagram showing how the MCL is determined for multiple material phase problems.

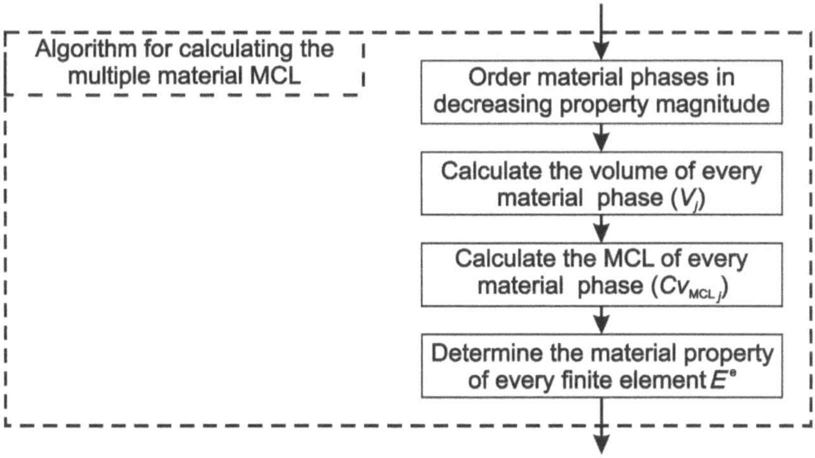

FIGURE 4.12 Schematic diagram showing the algorithm for calculating the MCL for multiple material phases problems.

$$E^e = \begin{cases} E_0 \text{ if } \xi^e = 0 \\ E_j \text{ otherwise} \end{cases}$$

$$E_0 = 10^{-6} \times E_{N_{MP}}$$

$$E_j = \begin{cases} E_1 & \text{if} & Cv_e > Cv_{MCL,1} \\ E_2 & \text{if} & Cv_{MCL,2} \leq Cv_e < Cv_{MCL,1} \\ \vdots & \vdots & \vdots \\ E_{N_{MP}} & \text{if} & Cv_{MCL,N_{MP}} \leq Cv_e < Cv_{MCL,N_{MP}-1} \end{cases} \tag{4.23}$$

where E^e is the Young's modulus of the eth FE; ξ^e is the fraction of the eth FE in the structure inside of the structural domain; E_0 is the Young's modulus of the void phase; $E_{N_{MP}}$ is the Young's modulus of the N_{MP} material (i.e., the weakest material); E_j is the Young's modulus of the jth material; Cv_e is the criterion value for the eth FE.

4.8 DETERMINATION OF THE STRUCTURAL SHAPE OR SURFACE

The boundary of a structure depends on the interaction of the MCL with the criterion distribution, which for 2D structures generates the MCL isoline and for 3D structures, it generates the MCL isosurface, Fig. 4.13. The following sections describe how the isolines and isosurfaces are extracted for the different structural applications.

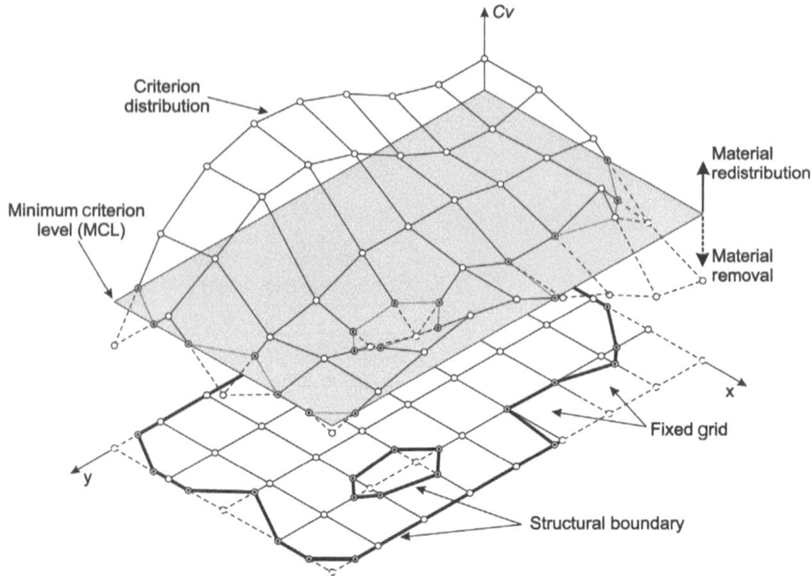

FIGURE 4.13 Structural boundary determined by the intersection of the MCL with the selected criterion distribution.

4.8.1 Determination of the Isolines for 2D Problems

To generate the MCL isoline, the Marching Triangles algorithm [34] was used. It considers each finite element independently as a triangular cell. In the case where the element has a quadrilateral shape, it is subdivided into four triangular cells by introducing a point at the centroid of that element. The criterion value of the central point is obtained by calculating the average values from the four corner nodes of the original element, or by interpolation using the values at the Gauss points of the element. The algorithm assumes that a contour can only pass through a triangular cell in one of two ways. This algorithm requires the value of the MCL together with the value of the criterion at each corner of the cell, and consists of two steps: (1) identify from Fig. 4.14, the topological state of each cell; and (2) determine the shape of the contour of the MCL isoline through each cell. The interaction of an isoline through a triangular cell can have a maximum of four different topological states, Fig. 4.14. The value (1) at a corner means that its criterion value is greater than the MCL whereas a value of (0) at a corner means that its criterion value is less than the MCL. When the corner in an edge of a cell has different values (0 and 1 or vice versa), this indicates that the MCL isoline intersects that edge, which is the case for the topological states of Fig. 4.14B and C.To find the intersection point in the two edges of the triangular cess, linear interpolation was used. The shape of the MCL isoline

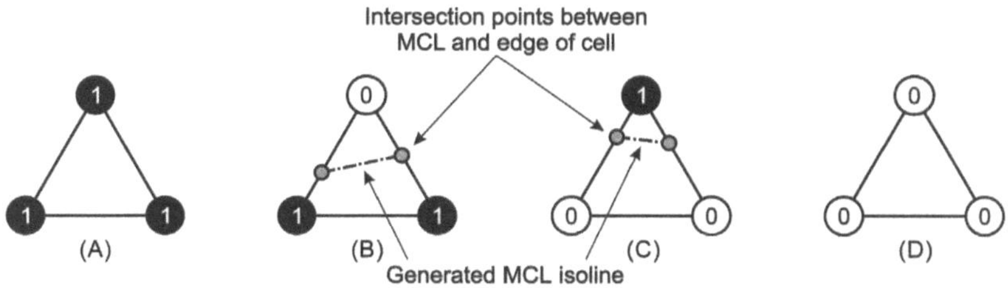

FIGURE 4.14 The four different topological states look-up table, used by the Marching Triangles algorithm to determine when an isoline intersects a triangular cell.

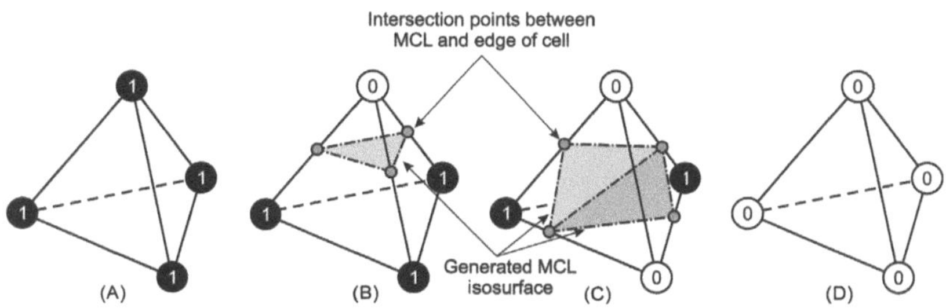

FIGURE 4.15 The four different topological states look-up table, used by the Marching Tetrahedra algorithm to determine when an isosurface intersects a tetrahedron element.

through the cell is then obtained by connecting these intersecting points between the opposite edges as shown in Fig. 4.14B and C. This process of boundary extraction is clearly shown in Fig. 4.13 for a 2D topology.

4.8.2 Determination of the Isosurfaces for 3D Problems

To generate the MCL isosurface for 3D problems, the Marching Tetrahedra algorithm [35] was used. A 3D structure can be represented using any combination of the following 3D finite elements: (A) tetrahedrons; (b) hexahedrons; and (c) pentahedrons. In the case where the FE is a tetrahedron, it is simply interpreted as a tetrahedron element. When the FE is a pentahedron it needs to be subdivided into 14 tetrahedrons, and when the FE is a hexahedron, it needs to be subdivided into 24 tetrahedrons. As for the marching triangle algorithm, the interaction of an isosurface through a tetrahedron can have a maximum of four different topological states, Fig. 4.15. The value (1) at a corner means that its criterion value is greater than the MCL whereas a value of (0) at a corner means that its criterion value is less than the MCL. The cases represented by Fig. 4.6B and C are the only two cases where triangulation is required. To find the intersection point between of the MCL with the edge of the element, linear interpolation was used. The shape of the

MCL isosurface through the element is then obtained by connecting these three intersecting points as shown in Fig. 4.6B and C.

4.9 STRUCTURAL BOUNDARY STABILIZATION

When the MCL is modified, the structural boundary changes and this affects the criterion distribution. Therefore, before the next iteration is started, an iterative process of reanalysis and material redistribution is carried out until the change in the domain volume between successive boundary adjustments is less than a minimum volume change limit ($\Delta V_{\mathrm{Min}_{\mathrm{Lim}}}$) or ($\% V_{\mathrm{Min}_{\mathrm{Lim}}}$) calculated using (Eq. 4.24 or 4.25).

$$\Delta V = \left| 1 - \frac{V_k}{V_{k-1}} \right| \tag{4.24}$$

$$\%\Delta V = \left| 1 - \frac{V_k}{V_{k-1}} \right| \times 100 \tag{4.25}$$

where (k) corresponds to the kth stabilization process iteration number.

REFERENCES

[1] C.Y. Lin, L.S. Chao, Automated image interpretation for integrated topology and shape optimization, Struct. Multidisc. Optim. 20 (2000) 125–137.

[2] R. Ansola, J. Canales, J.A. Tárrago, J. Rasmussen, An integrated approach for shape and topology optimization of shell structures, Comput. Struct. 80 (2002) 449–458.

[3] M.P. Bendsøe, N. Kikuchi, Generating optimal topologies in structural design using a homogenization method, Comput. Methods. Appl. Mech. Eng. 71 (1988) 197–224.

[4] B. Hassani, E. Hinton, Homogenization and Structural Topology Optimization: Theory, Practice and Software, Springer Verlag, 1999.

[5] D. Lee, S. Park, S. Shin, Node-wise topological shape optimum design for structural reinforced modeling of Michell-type concrete deep beams, J. Solid Mech. Mater. Eng. 1 (9) (2007) 1085–1096.

[6] K. Maute, E. Ramm, Adaptive topology optimization, Struct. Optim. 10 (1995) 100–112.

[7] Y.M. Xie, G.P. Steven, Evolutionary Structural Optimization, Springer, Berlin: Heidelberg. New York, 1997.

[8] L. Shao, H. Zhuo, Curve fitting with Bezier cubics, Graph. Models Image Process. 58 (3) (1996) 223–232.

[9] S.Y. Woon, L. Tong, O.M. Querin, G.P. Steven GP 2003: Optimising Topologies Through a Multi-GA System, WCSMO 5, Venice.

[10] H. Kim, M.J. Garcia, O.M. Querin, G.P. Steven, Y.M. Xie, Introduction of fixed grid in evolutionary structural optimisation, Eng. Comput. 17 (4) (2000) 427–439.

[11] H. Kim, O.M. Querin, G.P. Steven, Y.M. Xie, Improving efficiency of evolutionary structural optimization by implementing fixed grid mesh, Struct. Multidiscip. Optim. 24 (6) (2003) 441–448.

[12] C. Cui, H. Ohmori, M. Sasaki, Computational morphogenesis of 3D structures by extended ESO method, J. IASS 44 (141) (2003) 51–61.

[13] M. Victoria, P. Marti, O.M. Querin, Topology design of two-dimensional continuum structures using isolines, Comput. Struct. 87 (1–2) (2009) 101–109.

[14] M. Victoria, O.M. Querin, P. Martí, Topology design for multiple loading conditions of continuum structures using isolines and isosurfaces, Finite Elements Anal. Des. 46 (2010) 229–237.

[15] M. Victoria, O.M. Querin, P. Martí, Topology design of three-dimensional continuum structures using isosurfaces, Adv. Eng. Softw. 42 (2011) 671–679.

[16] M. Victoria, O.M. Querin, P. Martí, Generation of strut-and-tie models by topology design using different material properties in tension and compression, Struct. Multidiscip. Optim. 44 (2) (2011) 247–258.

[17] M. Victoria, O.M. Querin, C. Díaz, P. Martí, The effects of membrane thickness and asymmetry in the topology optimization of stiffeners for thin-shell structures, Eng. Optim. 46 (7) (2014) 880–894. Available from: http://dx.doi.org/10.1080/0305215X.2013.813023.

[18] O.M. Querin, M. Victoria, P. Martí, Topology optimization of truss-like continua with different material properties in tension and compression, Struct. Multidiscip. Optim. 42 (1) (2010) 25–32.

[19] O.M. Querin, M. Victoria, C. Díaz, P. Martí, Layout optimization of multi-material continuum structures with the isolines topology design method, Eng. Optim. 47 (2) (2015) 221–237. Available from: http://dx.doi.org/10.1080/0305215X.2014.882332.

[20] P. Pálfi, Locally orthotropic femur model, J. Comput. Appl. Mech. 5 (1) (2004) 103–115.

[21] P. Dewhurst, A general optimality criterion for combined strength and stiffness of dual-material-property structures, Int. J. Mech. Sci. 47 (2005) 293–302.

[22] A.U. Nair, Evolutionary numerical methods applied to minimum weight structural design and cardiac mechanics. PhD Thesis, Mechanical Engineering and Applied Mechanics, University of Rhode Island, 2005.

[23] D.G. Taggart, P. Dewhurst, A.U. Nair, System and method for finite element based topology optimization, WO/2007/076357 (Patent application), 2007.

[24] W.A. Jadayil, Revision of the recent heterogeneous solid object modeling techniques, Jordan J. Mech. Ind. Eng. 4 (6) (2011) 779–788.

[25] M.J. García-Ruiz, G.P. Steven, Fixed grid finite elements in elasticity problems, Engineering Computations 16 (2) (1999) 145–164.

[26] M.J. Garcia, O.E. Ruiz, L.M. Ruiz, O.M. Querin, Fixed grid finite element analysis for 3D linear elastic structures, Computational Mechanics (WCCM VI), China, 2004.

[27] M.J. Garcia, M. Henao, O.E. Ruiz, Fixed grid finite element analysis for 3D structural problems, Int. J. Comput. Methods 2 (4) (2005) 569–585.

[28] F.S. Maan, O.M. Querin, D.C. Barton, Extension of the fixed grid finite element method to eigenvalue problems, Adv. Eng. Softw. 38 (2007) 607–617.

[29] M.J. Garcia, G.P. Steven, Fixed grid finite element analysis in structural design and optimisation, in: Second ISSMO/AIAA Internet Conference on Approximations and Fast Reanalys is in Engineering Optimization, May 25–June 2, 2000.

[30] S.P. Timoshenko, J.N. Goodier, Theory of Elasticity, 3rd ed., McGraw-Hill, 1982, pp. 223–224. (section 77).

[31] E. Cervera, J. Trevelyan, Evolutionary structural optimisation based on boundary representation of NURBS. Part I: 2D algorithms, Comput. Struct. 83 (2005) 1902–1916.

[32] Y.M. Xie, G.P. Steven, Optimal design of multiple load case structures using an evolutionary procedure, Eng. Comput. 11 (1994) 295–305.

[33] V. Young, O.M. Querin, G.P. Steven, 3D and multiple load case bi-directional evolutionary structural optimization (BESO), Struct. Optim. 18 (1999) 183–192.

[34] A. Hilton, J. Illingworth, Technical Report CVSSP 01. Marching triangles: Delaunay implicit surface triangulation, University of Surrey, Guildford, UK, 1997

[35] G.M. Treece, R.W. Prager, A.H. Gee, Regularised marching tetrahedra: improved iso-surface extraction, Comput. Graph. 23 (4) (1999) 583–598.

Chapter 5

Hands-On Applications of Structural Optimization

5.1 INTRODUCTION

Over the past 100 years of structural optimization, there is a set of "classical" problems which have been used to test the validity of emerging topology optimization methods. However, more importantly, these examples are essential to show those starting to use topology optimization, the different characteristics that emerge within a structure when topology optimization is used to generate such a structure. The classical topology optimization problems are:

1. Michell cantilever;
2. Messerschmidt-Bölkow-Blohm (MBB) beam;
3. Michell cantilever with fixed circular boundary;
4. Michell beam with fixed supports;
5. Michell beam with roller supports;
6. Square under torsion;
7. Michell beam with roller support and multiple load cases;
8. Prager cantilever;
9. Inverter mechanism;
10. Gripper mechanism; and
11. Crunching mechanism.

Exact analytical solutions exist for some of these problems. A good means of comparing the numerical results with the exact analytical ones is by using the nondimensional mass of the structure given by Eq. (5.1) [1], or equivalently by Eq. (5.2) if the structure is made from a homogeneous material. A slight variation of these equations is given by Eqs. (5.3) and (5.4), proposed by Srithongchai and Dewhurst [2] for problems with different behavior in the tensile and compressive regions of the structure.

$$\overline{M} = M_{\text{Opt}} \frac{\sigma_e}{FL_{\text{nom}}\rho} \tag{5.1}$$

$$\overline{M} = V_{\text{Opt}} \frac{\sigma_e}{FL_{\text{nom}}} \tag{5.2}$$

Topology Design Methods for Structural Optimization.
DOI: http://dx.doi.org/10.1016/B978-0-08-100916-1.00005-2

$$\overline{M} = \frac{M_{\text{Opt}}}{FL_{\text{nom}}\left(\frac{\rho_T}{\sigma_T} + \frac{\rho_C}{\sigma_C}\right)} \tag{5.3}$$

$$\overline{M} = \frac{V_{\text{Opt}}}{FL_{\text{nom}}\left(\frac{1}{\sigma_T} + \frac{1}{\sigma_C}\right)} \tag{5.4}$$

where M_{Opt} and V_{Opt} are the optimal mass and volume of the structure, respectively; σ_e is the elastic limit of the structure; and σ_T, σ_C are the tensile and compressive stresses in the optimal structure, note that for numerical solutions these stresses can be the mean stress or the mean tensile and compressive stresses experienced by the structure; F is the applied load; L_{nom} is a specified nominal length; ρ is the density of the material; and ρ_T, ρ_C are the tensile and compressive material densities respectively, in some instances these last three density terms could also represent the cost per unit volume of the structure.

This chapter presents the 11 examples using the methods introduced in this book and shows the different possible solutions which can be obtained depending on the method used.

5.2 MICHELL CANTILEVER

The Michell cantilever [3] consists of a rectangular design domain with a built-in support at the left side and a single vertical applied load in the middle of the right end, Fig. 5.1. There are three classic Michell cantilevers which correspond to the aspect ratio ($\overline{L}$) of the length of the beam to its height, Eq. (5.5): (a) $\overline{L} = 0.5$, (b) $\overline{L} = 1.82196$, and (c) $\overline{L} = 3.35889$ [3]. For this problem the nominal length used is $L_{\text{nom}} = L$.

$$\overline{L} = L/h \tag{5.5}$$

where L is the length of the beam and h is the height of the beam.

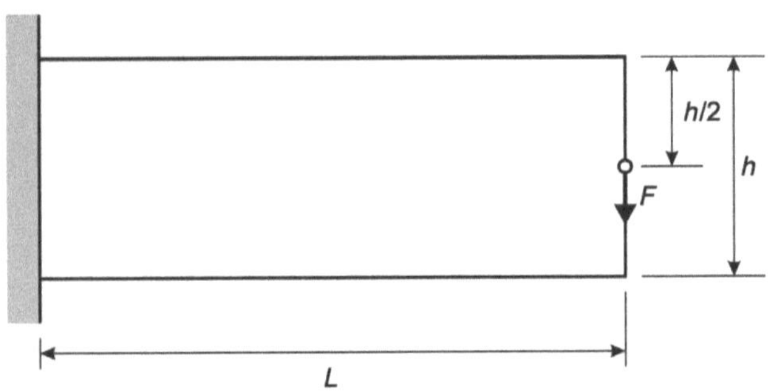

FIGURE 5.1 Michell cantilever.

TABLE 5.1 TTO Settings for the Three Cases of the Michell Cantilever

L/H	σ_c/σ_t	Load	Angle	Iteration	Grid	Symmetry
0.5	1	1	−90	1	50	Yes
1.82196	1	1	−90	190	50	Yes
3.35889	1	1	−90	200	50	Yes

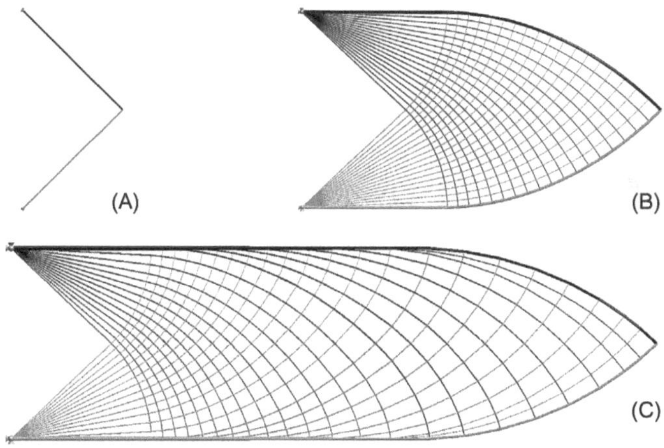

FIGURE 5.2 The optimal topologies for the three cases of the Michell Cantilever: (A) $\overline{L}=0.5$, (B) $\overline{L}=1.82196$, and (C) $\overline{L}=3.35889$.

For these three cases, the analytically derived optimal nondimensional masses are: (a) $\overline{M}=1$, (b) $\overline{M}=6.07386$, and (c) $\overline{M}=16.44137$ [3]. The Growth method of Chapter 3, Discrete Method of Structural Optimization, was used to solve the three cases using the included program TTO. The settings used by the program for the three problems are given in Table 5.1.

The optimal topology for the three cases is shown in Fig. 5.2, where the tensile bars are in red (black in print versions) and the compressive bars in blue (grey in print versions). The summary of the number of joints, bars, and nondimensional masses is given in Table 5.2. The topologies which emerge for cases (b) and (c) show that the topologies are made up of two fan like regions consisting of straight radially positioned bars and one large region consisting of a Hencky net [4].

5.3 MESSERSCHMIDT-BÖLKOW-BLOHM BEAM

The Messerschmidt-Bölkow-Blohm (MBB) beam [3] is the floor of the passenger airliner Airbus. The design domain is a rectangle of length $2L$ and height h, Fig. 5.3. The two most prominent cases correspond to the following two aspect ratios ($\overline{L}$) of the half-length of the beam to its height (Eq. 5.5): (a) $\overline{L} = 2.40196$ and (b) $\overline{L} = 5.49846$. For this problem, the nominal length used is $L_{\mathrm{nom}} = h$.

The analytically derived optimal nondimensional masses are: (a) $\overline{M} = 9.95569$ and (b) $\overline{M} = 39.25304$ [5]. The Growth method of Chapter 3, Discrete Method of Structural Optimization, was used to solve this problem. The settings used by the TTO program are given in Table 5.3. Only half of the beam was used for this analysis by applying symmetry boundary conditions about the Y axis, which meant that all nodes on the left-hand edge were only allowed a vertical displacement.

TABLE 5.2 Optimal Parameters for the Three Cases of the Michell Cantilever

L/H	Joints	Bars	$\overline{M}$
0.5	3	2	1
1.82196	363	722	6.0749433
3.35889	377	750	16.4504388

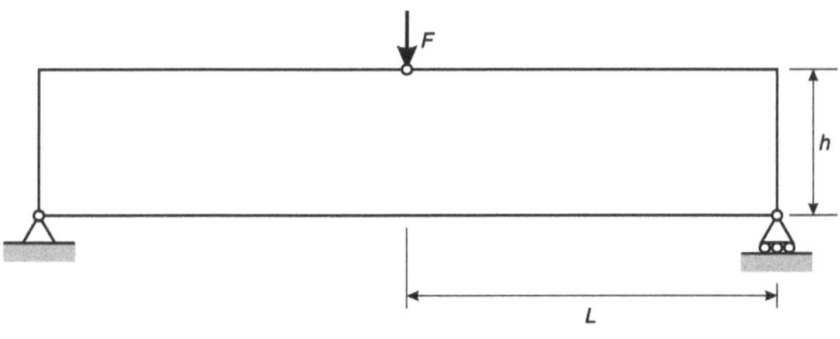

FIGURE 5.3 MBB beam design domain.

TABLE 5.3 TTO Settings for the Two Cases of the MBB Beam

L/H	σ_c/σ_t	Load	Angle	Iteration	Grid	Symmetry
2.40196	1	1	−90	292	50	No
5.49846	1	1	−90	299	50	No

The optimal topology for the two cases is shown in Fig. 5.4, where the tensile bars are in red (black in print versions) and the compressive bars in blue (dark grey in print versions). There is a bar in green (light grey in print versions) in Fig. 5.4A and B, the green color means that the bar has no cross-sectional area but is required to keep the structure statically determinate. The summary of the number of joints, bars and nondimensional masses are given in Table 5.4. The same fan like region consisting of straight radially positioned bars and one large region consisting of a Hencky net are also observed in these topologies [4].

The ITD method of Chapter 4, Continuous Method of Structural Optimization, was also used to solve this problem using the included program liteITD. The dimensions of the model were: $L = 360.294$ mm, $h = 150$ mm, thickness = 1 mm. As half the structure was modeled using symmetry conditions, the applied load was $F = 500$ N, Fig. 5.5. The mesh used was of 40×96 FE. This means that the elements are not exactly square as their vertical size was 3.75 and the horizontal size was 3.7530625 mm. The modulus of elasticity used for I elements was 210×10^3 MPa and the Poisson's ratio was 0.3. For liteITD, the modulus of elasticity of O elements

(A)
(B)

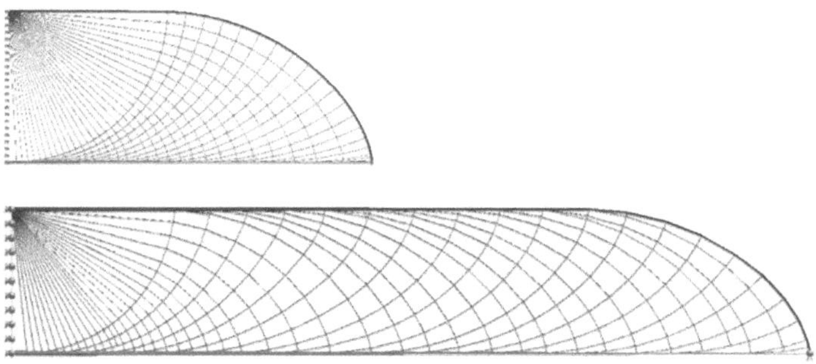

FIGURE 5.4 The optimal topologies for the two cases of the MBB: (A) $\overline{L} = 2.40196$ and (B) $\overline{L} = 5.49846$.

TABLE 5.4 Optimal Parameters for the Two Cases of the Michell Cantilever

L/H	Joints	Bars	$\overline{M}$
2.40196	315	606	9.95998711
5.49846	300	588	39.2925984

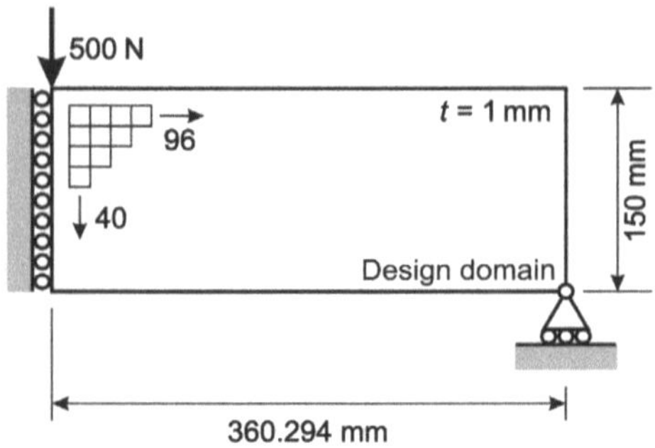

FIGURE 5.5 MBB beam design domain for use in ITD.

FIGURE 5.6 MBB beam design for a volume fraction of 58.7%.

TABLE 5.5 MBB Results

(V/V_0)	(D_{Max})	(σ_{vM}^{max})	(σ_{vM}^{mean})	(σ_{vM}^{min})
0.587	−0.213	10.68	24.96	360.39

was set to be 10^{-4} times smaller than for I elements, so for this problem it was 21 MPa.

The emergent topology for a volume fraction of 58.7% is shown in Fig. 5.6. The maximum vertical displacement (D_{Max}) in mm; maximum (σ_{vM}^{max}), mean (σ_{vM}^{mean}), and minimum (σ_{vM}^{min}) von Mises stresses in MPa are given in Table 5.5. Its resulting nondimensional mass was $\overline{M} = 9.36168$.

5.4 MICHELL CANTILEVER WITH FIXED CIRCULAR BOUNDARY

The Michell cantilever with fixed circular boundary consists of the rectangular design domain of Fig. 5.7, with a length equal to $L + 1.5r$ and with a

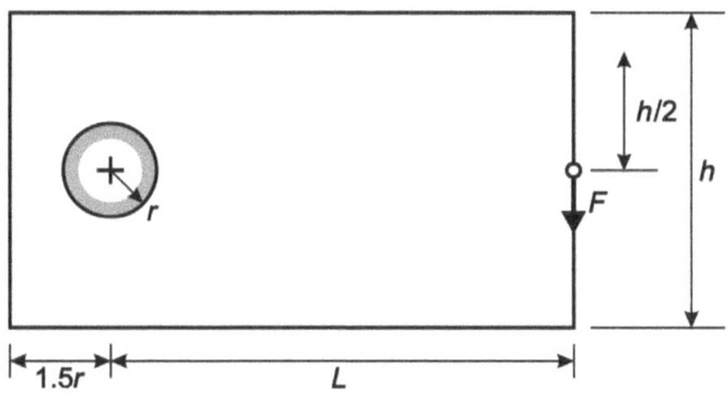

FIGURE 5.7 Design domain for the Michell cantilever with fixed circular boundary.

TABLE 5.6 TTO Settings for the Two Cases of the Michell Cantilever With Fixed Circular Boundary

L/r	σ_c/σ_t	Fy	α	Iteration	Grid	Symmetry
10	1	−1	−90	210	50	Yes
15	1	−1	−90	210	50	Yes

vertically applied force at the center of the right end. The height is given by Eq. (5.6) and the nominal length used is $L_{nom} = L$. The optimal analytical nondimensional mass is given by Eq. (5.7) [3,6].

$$h = \frac{\sqrt{2}}{e^{\pi/4}} L = 0.6447939L \tag{5.6}$$

$$\overline{M} = 2 \text{ Ln}\left(\frac{L}{r}\right) \tag{5.7}$$

where r is the radius of the supporting region.

The Growth method of Chapter 3, Discrete Method of Structural Optimization, was used to solve this problem. The settings used by the TTO program are given in Table 5.6 for the two cases studied: (a) $L/r = 10$ and (b) $L/r = 15$. The analytical optimum nondimensional masses are for (a) $\overline{M} = 4.60517$ and for (b) $\overline{M} = 5.41610$.

The optimal topology for the two cases is shown in Fig. 5.8, where the tensile bars are in red (black in print versions) and the compressive bars in blue (gray in print versions). These topologies consist of only a Hencky net [4]. The summary of the number of joints, bars, and nondimensional masses are given in Table 5.7.

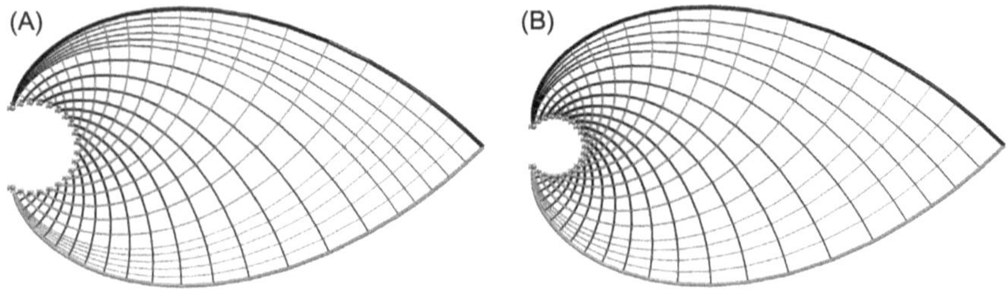

FIGURE 5.8 Optimal design for the two cases of the Michell cantilever with fixed circular boundary: (A) $L/r = 10$ and (B) $L/r = 15$.

TABLE 5.7 Optimal Parameters for the Two Cases of the Michell Cantilever With Fixed Circular Boundary

L/r	Joints	Bars	$\overline{M}$
10	425	810	4.62181642
15	419	796	5.43593058

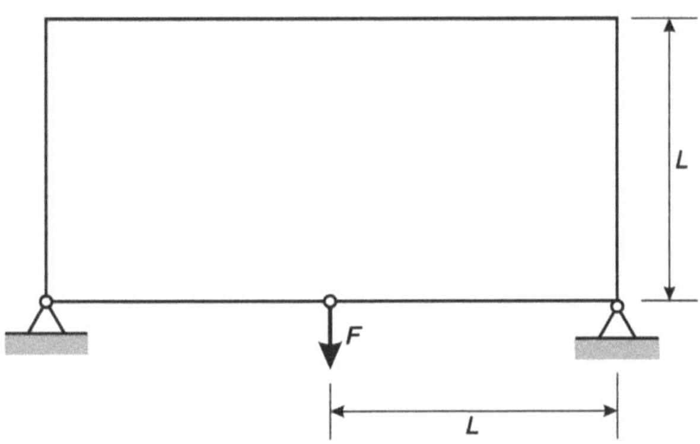

FIGURE 5.9 Design domain for the Michell beam with fixed supports.

5.5 MICHELL BEAM WITH FIXED SUPPORTS

The Michell beam with fixed supports [6] consists of a rectangular design domain of width 2 L and height L, the nominal length used is $L_{\text{nom}} = L$. It has two fixed supports at the bottom left and right corner points, and a single vertical applied load in the middle of the bottom edge, Fig. 5.9.

The optimal nondimensional mass of this beam for the case of equal stress limits for the beams in tension and compression is

$\overline{M} = 1 + \frac{\pi}{2} = 2.570796$ [6]. The Growth method of Chapter 3, Discrete Method of Structural Optimization, was used to solve this problem. The settings used by the TTO program are given in Table 5.8. The only added setting to the model was to prevent horizontal displacement at the point where the force was applied, as otherwise the structure would become a mechanism.

The optimal topology consists of a 90 degree fan like region emerging from the point of application of the load and consisting of straight radially positioned bars, Fig. 5.10, where the tensile bars are in red (black in print versions) and the compressive bars in blue (grey in print versions). The summary of the number of joints, bars and nondimensional masses are given in Table 5.9.

The ITD method of Chapter 4, Continuous Method of Structural Optimization, was also used to solve this problem. The dimensions of the model were: $L = 100$ mm, thickness = 1 mm, the applied load was $F = 1000$ N and the mesh used was of 200 × 100 FE. The modulus of elasticity used for I elements was 210×10^3 MPa, the Poisson's ratio was 0.3, and the modulus of elasticity of O elements was 21 MPa.

TABLE 5.8 TTO Settings for the Michell Beam With Fixed Supports

L/H	σ_c/σ_t	Load	Angle	Iteration	Grid	Symmetry
2	1	−1	−90	150	50	Yes

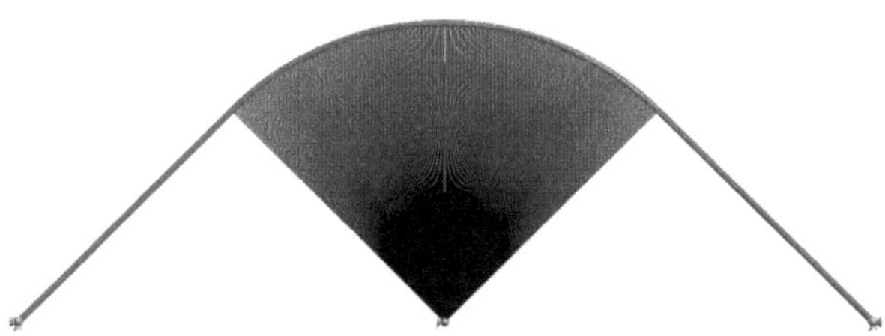

FIGURE 5.10 Optimal design of the Michell beam with fixed supports.

TABLE 5.9 Optimal Parameters for the Michell Beam With Fixed Supports

L/H	Joints	Bars	$\overline{M}$
2	302	599	2.57079994

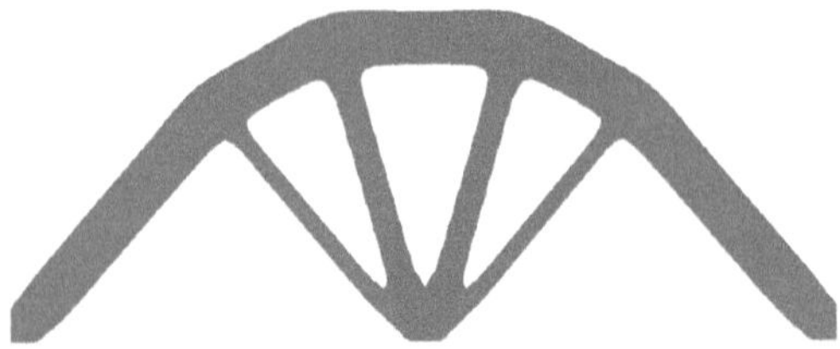

FIGURE 5.11 Michell beam with fixed supports design for a volume fraction of 20.68%.

TABLE 5.10 Michell Beam With Fixed Supports Results

(V/V_0)	(D_{Max})	(σ_{vM}^{max})	(σ_{vM}^{mean})	(σ_{vM}^{min})
0.21	−0.02	729.16	57.70	29.06

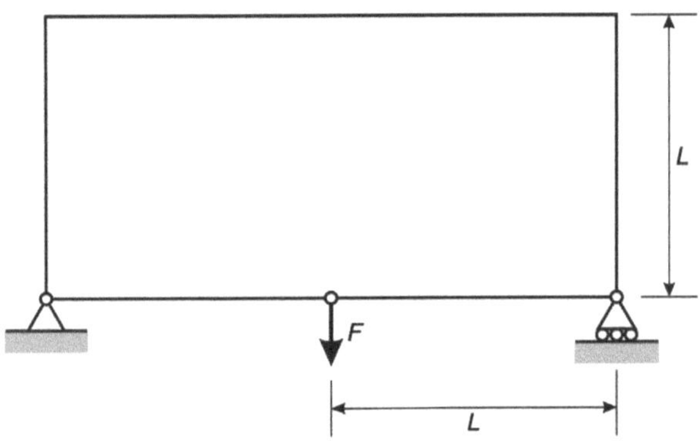

FIGURE 5.12 Design domain for the Michell beam with one roller support.

The resulting topology for a volume fraction of 20.68% is shown in Fig. 5.11. The maximum vertical displacement (D_{Max}) in mm; maximum (σ_{vM}^{max}), mean (σ_{vM}^{mean}), and minimum (σ_{vM}^{min}) von Mises stresses in MPa are given in Table 5.10. Its resulting nondimensional mass was $\overline{M} = 2.38641$.

5.6 MICHELL BEAM WITH ROLLER SUPPORT

The Michell beam with a roller support [6] consists of a rectangular design domain of width 2 L and height L, the nominal length used is $L_{nom} = L$. It has one fixed supports at the bottom left-hand corner and one roller support at the right corner, with a single vertical applied load in the middle of the bottom edge, Fig. 5.12.

The optimal nondimensional mass of this beam for the case of equal stress limits for the beams in tension and compression is $\overline{M} = \pi = 3.14159$ [6]. The Growth method of Chapter 3, Discrete Method of Structural Optimization, was used to solve this problem, the settings used by the TTO program are given in Table 5.11.

The optimal topology consists of a hemispherical shape made from a fan like region emerging from the point of application of the load consisting of straight radially positioned bars, Fig. 5.13, where the tensile bars are in red (black in print versions) and the compressive bars in blue (grey in print versions). The summary of the number of joints, bars, and nondimensional masses are given in Table 5.12.

The ITD method of Chapter 4, Continuous Method of Structural Optimization, was also used to solve this problem. The dimensions of the model were: $L = 100$ mm, thickness = 1 mm, as half the structure was modeled using symmetry conditions, the applied load was $F = 500$ N and the mesh used was of 100×100 FE, Fig. 5.14. The modulus of elasticity used

TABLE 5.11 TTO Settings for the Michell Beam With One Roller Support

L/H	σ_c/σ_t	Load	Angle	Iteration	Grid	Symmetry
2	1	1	−90	160	50	No

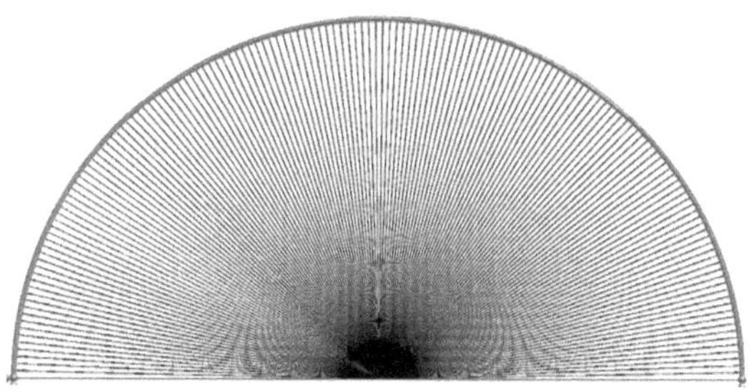

FIGURE 5.13 Optimal design of the Michell beam with one roller support.

TABLE 5.12 Optimal Parameters for the Michell Beam With One Roller Support

L/H	Joints	Bars	$\overline{M}$
2	160	317	3.14169618

for I elements was 210×10^3 MPa, the Poisson's ratio was 0.3, and the modulus of elasticity of O elements was 21 MPa.

The resulting topology for a volume fraction of 29.66% is shown in Fig. 5.15. The maximum vertical displacement (D_{Max}) in mm; maximum (σ_{vM}^{max}), mean (σ_{vM}^{mean}), and minimum (σ_{vM}^{min}) von Mises stresses in MPa are given in Table 5.13. Its resulting nondimensional mass was $\overline{M} = 2.81462$.

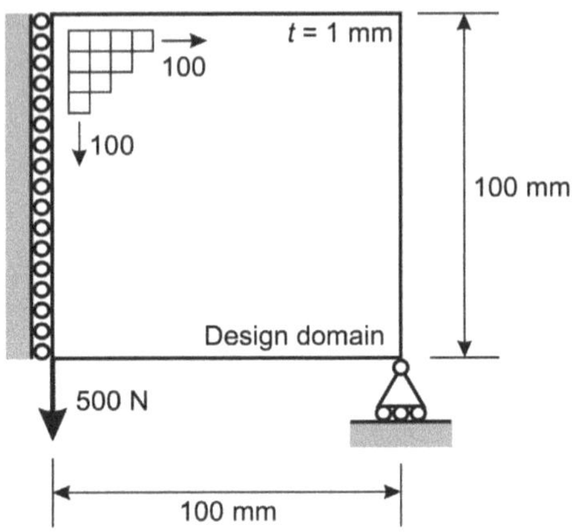

FIGURE 5.14 Design domain for Michell beam with one roller support for use in ITD.

FIGURE 5.15 Michell beam with fixed supports design for a volume fraction of 29.66%, showing the full design space.

TABLE 5.13 Michell Beam With Fixed Supports Results

(V/V_0)	(D_{Max})	(σ_{vM}^{max})	(σ_{vM}^{mean})	(σ_{vM}^{min})
0.30	−0.08	665.17	43.13	22.47

5.7 SQUARE UNDER TORSION

The square under torsion problem was introduced by Diaz and Bendsøe [7] and consists of a square design domain of dimensions 1000 mm × 1000 mm and thickness 100 mm, with a centrally located square hole of dimensions 200 mm × 200 mm where all the displacements are restricted, Fig. 5.16. The structure is subjected to two different load cases, represented by loads F_1 and F_2, each of magnitude 10 kN. Load case 1, represented by forces F_1 produces a counter clockwise torque of magnitude 20 kNm, and load case 2, represented by forces F_2, produces a clockwise torque of magnitude 20 kNm.

The ITD method of Chapter 4, Continuous Method of Structural Optimization, was used to solve this problem. The modulus of elasticity of I elements was 210×10^3 MPa, the Poisson's ratio was 0.3, and the modulus of elasticity of O elements was 21 MPa. The mesh used consisted of 250 × 250 four-node plane stress quadrilateral finite element with four Gauss integration points [8]. The design criteria used was the von Mises stress and the ITD parameters used are given in Table 5.14.

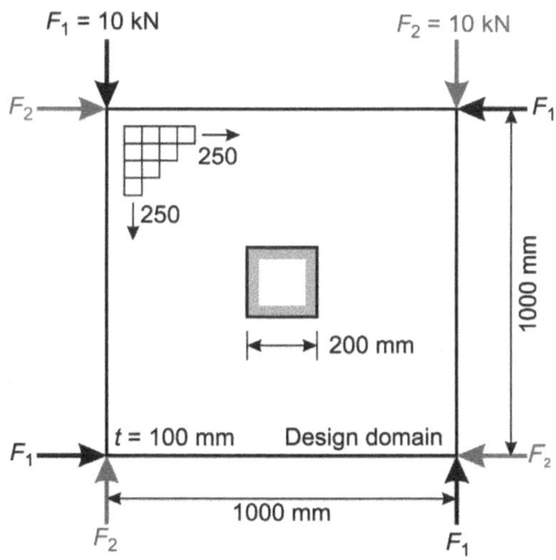

FIGURE 5.16 Design domain for the square under torsion.

TABLE 5.14 Square Under Torsion ITD Used Parameters

Total Iterations, N_{Tot}	Final Volume Fraction	Volume Change Limit, $\%\Delta V_{Min_{Lim}}$	Total Number of Load Cases, N_{LC}
50	30%	1.5	2

FIGURE 5.17 Final design for the square under torsion.

The design generated is shown in Fig. 5.17. It has a central region surrounding the supporting square which has a basic Hencky net arrangement, followed by two single bar elements connecting this shape to the location of the applied loads. This shape is in excellent agreement with that from Diaz and Bendsøe [7].

5.8 MICHELL BEAM WITH ROLLER SUPPORT AND MULTIPLE LOAD CASES

The design domain for the Michell beam with roller support and multiple load cases consists of a rectangular area of size $10\,\mathrm{m} \times 3\,\mathrm{m}$ and thickness = 0.1 m [7,9]. The structure has a fixed support in the bottom left-hand corner and a roller support in the bottom right-hand corner. It is applied with three identical downward loads F_1, F_2, and F_3, which are equally spaced at the bottom edge, Fig. 5.18. Two cases were considered: (a) when all three loads act at the same time consisting of a single load case and (b) when each load acts individually and a multiple load case condition exists with three distinctive load cases. The ITD method of Chapter 4, Continuous Method of Structural Optimization, was used to solve this problem. The model used a mesh of 160×48 four-node plane stress quadrilateral finite elements with four Gauss integration points [8]. The modulus of elasticity of I elements was 210×10^3 MPa, the Poisson's ratio was 0.3, and the modulus of elasticity of O elements was 21 MPa. The design criteria used was the von Mises stress and the ITD parameters used are given in Table 5.15.

The resulting design for case (a), Fig. 5.19A, produces a truss-like structure that is not rigid. Hence for this problem, the single load case arrangement produces an unstable structure based on squared frames. The resulting design for case (b), Fig. 5.19B, produces a stable structure based on triangular frames, in agreement with the results of Diaz and Bendsøe [7] and Bendsøe and Sigmund [9].

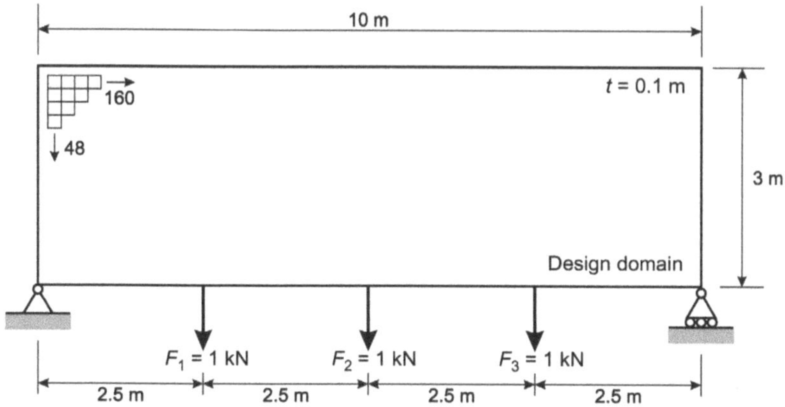

FIGURE 5.18 Design domain for the multiply loaded Michell beam with roller support.

TABLE 5.15 ITD Parameters for Multiply Loaded Michell Beam With Roller Support

Total Iterations, N_{Tot}	Final Volume Fraction	Volume Change Limit, $\%\Delta V_{Min_{Lim}}$	Total Number of Load Cases, N_{LC}
50	20%	1.5	3

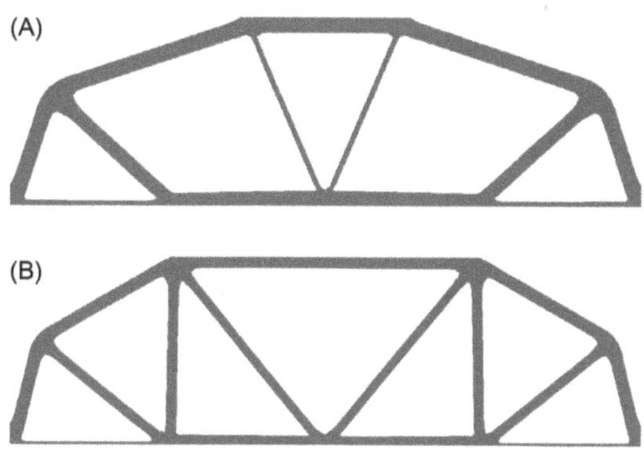

FIGURE 5.19 Final designs for the multiply loaded Michell beam with roller support. (A) Design when all loads act simultaneously; (B) design for multiple load cases.

5.9 PRAGER CANTILEVER

The Prager Cantilever example was originally proposed by Prager [10] as a challenge to the reader to solve its optimum characteristics. The original boundary layout proposed by Prager is shown as the dashed rectangle in Fig. 5.20.

In order to capture accurately the angles of the emerging truss elements in the topology, such that these are properly aligned between the supports and load, a border around the true design domain was placed denoted by lengths $B_{1,2,3,4}$. For this problem the nominal length used is $L_{nom} = L$.

The optimal nondimensional analytical mass obtained by Srithongchai and Dewhurst [2] was $\overline{M} = 2.1513$. The Growth method of Chapter 3, Discrete Method of Structural Optimization, was used to solve this problem. The dimensions used were $h = 1$ and $L = 3$, represented by the dashed rectangle. The lengths of the border were $B_1 = B_2 = 0.5$, $B_3 = B_4 = 0$. The settings used by the TTO program are given in Table 5.16.

The optimal topologies are shown in Fig. 5.21 with the summary of the number of joints, bars, and nondimensional masses given in Table 5.17.

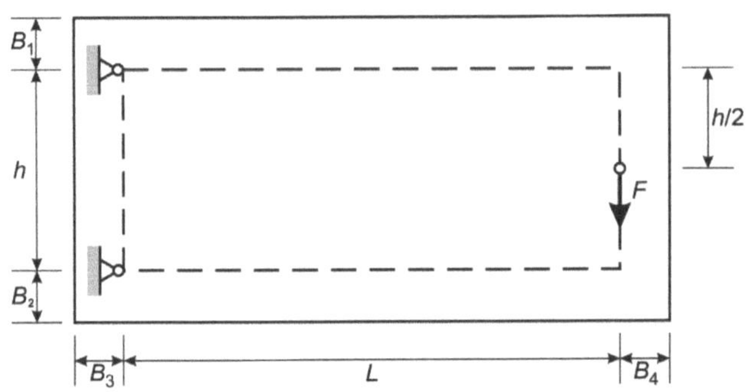

FIGURE 5.20 Design domain for the Prager cantilever.

TABLE 5.16 TTO Settings for the Two Models of the Prager Cantilever

L/H	σ_c/σ_t	Load	Angle	Iteration	Grid	Symmetry
1.5	0.5	1	−90	12 and 108	18	No

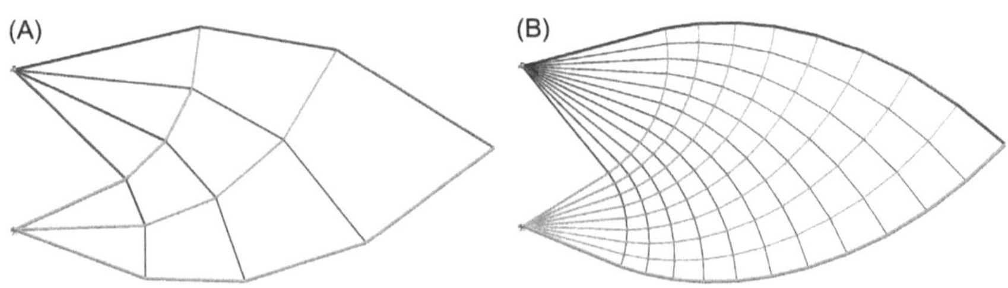

FIGURE 5.21 Final design for the Prager cantilever: (A) after 12 iterations and (B) after 108 iterations.

TABLE 5.17 Optimal Parameters for the Prager Cantilever

Iteration	Joints	Bars	$\overline{M}$
12	14	24	2.18310467
108	110	216	2.15319261

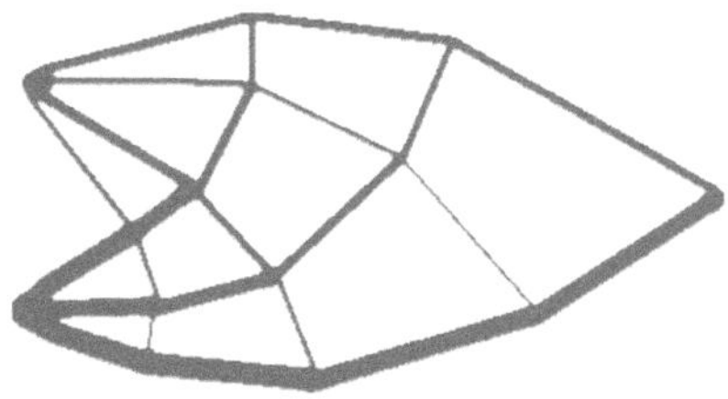

FIGURE 5.22 Final design for the Prager cantilever.

The ITD method of Chapter 4, Continuous Method of Structural Optimization, was also used to solve this problem. In order to use a regular mesh and looking at the topology generated by the TTO program, the dimensions of the borders were modified from those used for the TTO solution. The dimensions used were: $h = 1000$ mm, $L = 3000$ mm, $B_1 = 310.0285$ mm, $B_2 = 355.3001$ mm, $B_3 = 77.4571$ mm, $B_4 = 38.7285$ mm, with the thickness of the domain set at $t = 1$ mm. The entire design domain (solid outer rectangle) was subdivided with a mesh of 215×402 four-node plane stress quadrilateral element with four Gauss integration points [8]. The magnitude of the vertical applied force was $F = 10$ kN. The prescribed constant ratio between stresses was $k_1 = 2$. The modulus of elasticity was 210×10^3 MPa. Although Rozvany [11] suggested the use of a Poisson's ratio value of zero in order to obtain structures close to those of Michell, problems were experienced when this was tried with the orthotropic elements, so a Poisson's ratio of 0.3 was used. The optimal topology is shown in Fig. 5.22 which is in good agreement with Fig. 5.21A. Its resulting nondimensional mass was $\overline{M} = 2.0928$.

5.10 INVERTER MECHANISM

The inverter mechanism is used to change the direction of an actuating displacement or applied force. The design domain consists of a square of size 200 mm × 200 mm subdivided using a mesh of 100 × 100 square four-node finite elements, the thickness of the domain was $t = 1$ mm, Fig. 5.23. The input force was $F_{in} = 1$ N and the stiffness ratio was $k_{out}/k_{in} = 1$. The limit volume fraction (V^{Lim}) was 0.4 and the filter radius (r_{min}) was 6 mm.

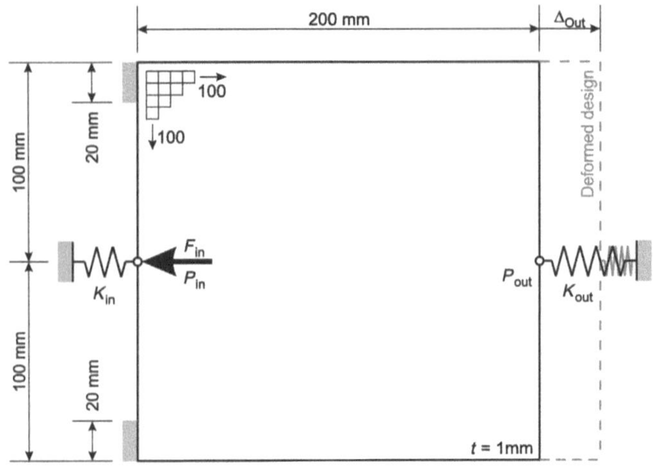

FIGURE 5.23 Design domain for the inverter mechanism.

FIGURE 5.24 Final design for the inverter mechanism.

The modulus of elasticity was 200×10^3 MPa, the Poisson's ratio was 0.3 and the density of the virtual material was $\rho_{\min} = 10^{-4}$, which is equivalent to 0.01% of the stiffness of a real material.

The SERA method of Chapter 3, Discrete Method of Structural Optimization, was applied to this problem to generate the resulting topology of Fig. 5.24, which is in good agreement with the optimum topologies of Sigmund [12] and Ansola et al. [13].

5.11 GRIPPER MECHANISM

The gripper mechanism converts a horizontally applied input force at the middle of the left side into a closing of the jaws at the opposite side of where

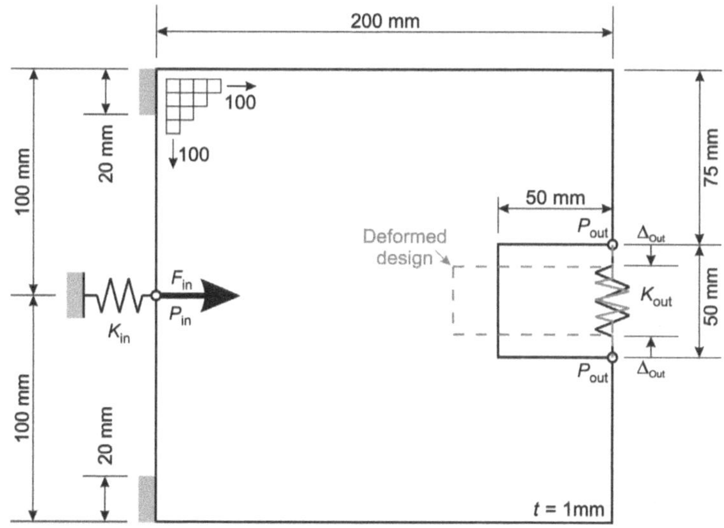

FIGURE 5.25 Design domain for the gripper mechanism.

the force is applied [12]. The design domain consists of a square of size 200 mm × 200 mm subdivided using a mesh of 100 × 100 square four-node finite elements, the thickness of the domain was $t = 1$ mm, Fig. 5.25. A square section of size 50 mm × 50 mm is removed from the center of the right side of the design domain to allow the mechanism to grip the workpiece, modeled by the output spring K_{out}.

The input force was $F_{in} = 1$ N with a stiffness ratio of $k_{out}/k_{in} = 1$. The limit volume fraction (V^{Lim}) was 0.2 and the filter radius (r_{min}) was 6 mm. The modulus of elasticity was 200×10^3 MPa, the Poisson's ratio was 0.3 and the density of the virtual material was $\rho_{min} = 10^{-4}$, which is equivalent to 0.01% of the stiffness of a real material.

The SERA method of Chapter 3, Discrete Method of Structural Optimization, was applied to this problem to generate the resulting topology of Fig. 5.26, which is in good agreement with the optimum topologies obtained by Sigmund [12].

5.12 CRUNCHING MECHANISM

The crunching mechanism converts a compressive "crunching" input force acting vertically on the right side of the design domain, into an outward displacement in the middle of the left edge [12]. The design domain consists of a square of size 200 mm × 200 mm subdivided using a mesh of 100 × 100 square four-node finite elements, the thickness of the domain was $t = 1$ mm, Fig. 5.27.

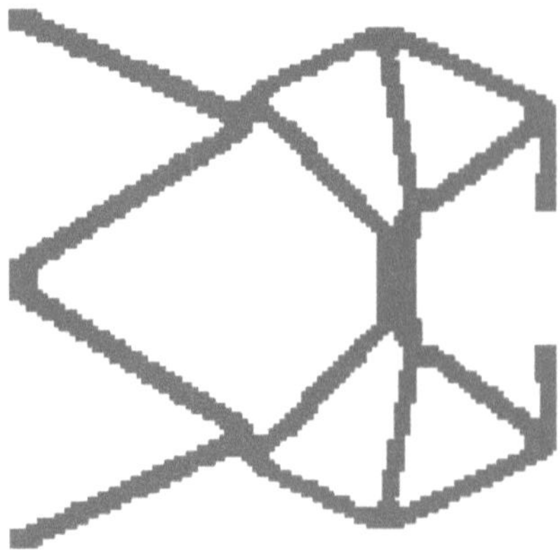

FIGURE 5.26 Final design for the gripper mechanism.

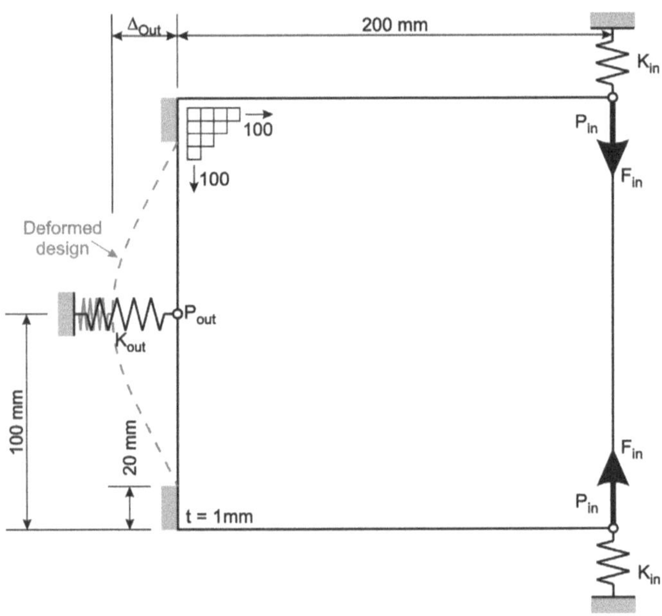

FIGURE 5.27 Design domain for the crunching mechanism.

The input force was $F_{in} = 1$ N with a stiffness ratio of $k_{out}/k_{in} = 1$. The limit volume fraction (V^{Lim}) was 0.4 and the filter radius (r_{min}) was 6 mm. The modulus of elasticity was 200×10^3 MPa, the Poisson's ratio was 0.3 and the density of the virtual material was $\rho_{min} = 10^{-4}$, which is equivalent to 0.01% of the stiffness of a real material.

The SERA method of Chapter 3, Discrete Method of Structural Optimization, was applied to this problem to generate the resulting topology

FIGURE 5.28 Final design for the crunching mechanism.

of Fig. 5.28, which is in good agreement with the optimum topologies obtained by Sigmund [12].

REFERENCES

[1] T. Lewiński, M. Zhou, G.I.N. Rozvany, Extended exact solution for least-weight truss layout. Part I: cantilever with a horizontal axis of symmetry, Int. J. Mech. Sci. 36 (1994) 375–398.

[2] S. Srithongchai, P. Dewhurst, Comparisons of optimality criteria for minimum-weight dual material structures, Int. J. Mech. Sci. 45 (2003) 1781–1797.

[3] G.I.N. Rozvany, Exact analytical solutions for some popular benchmark problems in topology optimization, Struct. Optim. 15 (1998) 42–48.

[4] M. Zhou, G.I.N. Rozvany, DCOC: an optimality criteria method for large systems. Part II: algorithm, Struct. Optim. 6 (1993) 250–262.

[5] T. Lewiński, M. Zhou, G.I.N. Rozvany, Extended exact solution for least-weight truss layout. Part II: unsymmetric cantilevers, Int. J. Mech. Sci. 36 (1994) 399–419.

[6] W.S. Hemp, Optimum Structures, Clarendon Press, Oxford, 1973.

[7] A.R. Diaz, M.P. Bendsøe, Shape optimization of structures for multiple loading conditions using a homogenization method, Struct. Optim. 4 (1992) 17–22.

[8] O.C. Zienkiewicz, R.L. Taylor, J.Z. Zhu, The Finite Element Method, Its Basis & Fundamentals, Elsevier Butterworth Heinemann, 2005.

[9] M.P. Bendsøe, O. Sigmund, Topology Optimization. Theory, Methods and Applications, Springer Verlag, 2003.

[10] W. Prager, A problem of optimal design, in: Proceedings of the Union of Theoretical and Applied Mechanics, Warsaw, 1958.

[11] G.I.N. Rozvany, A critical review of established methods of structural topology optimization, Struct. Multidiscip. Optim. 37 (2009) 217–237.

[12] O. Sigmund, On the design of compliant mechanisms using topology optimization, Mech. Struct. Mach. 25 (4) (1997) 493–524.

[13] R. Ansola, E. Vegueria, J. Canales, J. Tárrago, A simple evolutionary topology optimization procedure for compliant mechanism design, Finite. Elem. Anal. Des. 44 (2007) 53–62.

Chapter 6

Topology Optimization as a Digital Design Tool

6.1 INTRODUCTION

There has been an emergence of topology optimization as a very powerful and useful tool for the design of structures. This is self-evident by the implementation of topology and other forms of optimization in most commercial structural analysis programs such as ABAQUS [1], ANSYS [2], FEMtools [3], and MSC NASTRAN [4]. There has also been topology optimization specific software with either inbuilt or links to external analysis tools, such as OPTISTRUCT [5] and CATOPO [6]. As well as topology optimization tools available over the internet such as TOPOPT [7] and CalculiX [8]. Although such software is becoming easier to use, it still requires some knowledge of optimization or finite element analysis, which is essential for engineering structural design but may not be for general product design.

One of the reasons for the growth in the development of topology optimization has been the growth in computational power and speed. This too has had a remarkable impact to the design process, shifting it from the traditional paper-based concept of design to digital design. The theory of Digital Design has changed the concept of *form* into the concept of *formation* [9]. Instead of the graphical manipulation of a form or structure, a topology is formed or developed by the manipulation of the parametric representation of the design due to its overall behavior.

According to Oxman [10], there are five classes of digital design models: (1) Computer Aided Design (CAD); (2) Formation; (3) Generative; (4) Performance; and (5) Integrated compound models. The model which is related to the process of structural optimization is the Performance model, since the form (structure) is generated according to the desired performance or behavior of the object. In Ref. [10], this model is also subdivided into two subclasses: performance-based formation and performance-based generation models of design.

In the Performance-based formation model, the digital simulation of the external forces drives the formation process of the design. These types of designs use analytical simulation techniques that produce detailed parametric expressions of performance, which in turn can produce formation responses

Topology Design Methods for Structural Optimization.
DOI: http://dx.doi.org/10.1016/B978-0-08-100916-1.00006-4

to complex classes of performance requirements. Some of which can be acoustic, aerodynamic, cultural, ecological, environmental, financial cost, spatial, social, or structural.

In the Performance-based generation model, the performance simulations of the design drive the generation and/or formation processes in order to generate the form. The external forces can include acoustics, environmental, information, structural loads, and transportation.

Although structural optimization software is available, this functionality has not yet been extended to CAD software used by designers. This was pointed out by Kolarevic [11] who described the inadequacy of existing analytical CAD software in conceptual design. Kolarevic [12] lists several digital approaches to architectural design based on computational concepts and suggests that new categories could be incorporated as new processes become available, citing as an example structural performance-based generation and transformation of forms. Oxman [9] argues that current digital tools used for simulation, analysis, and evaluation of performance aspects do not currently provide generative and modification capabilities. The implementation of optimization tools together with structural analysis digital tools has finally allowed this to happen and is currently available for performance-based digital design.

A brief review of structural optimization applied to engineering design can be found in Kim et al. [13]. In Sigmund and Bendsøe [14], the versatility of applying topology optimization from the nanoscale to the size of aircraft components is shown. In Krog et al. [15], optimization is applied to the design of aircraft components. In Xie et al. [16], topology optimization was applied to architectural structures by duplicating some of the designs of Gaudí. Poncelet et al. [17] showed how topology optimization could be used as a tool for industrial design by applying it to the design of the fixtures for a satellite mirror, street lighting device, and an aircraft engine pylon. In Lee et al. [18], topology optimization was applied to automotive components. Wu et al. [19] showed its applicability to the drug manufacturing industry by designing a stent platform with drug reservoirs. These are but a few of the many studies in the application of topology optimization for engineering design. More recently, some papers have appeared where topology optimization has been applied to architectural and everyday products [20,21].

In order to show how topology optimization can be used as a digital tool to design a structure or a component, the following nine examples are investigated in this chapter:

1. Michell cantilever;
2. Tap or faucet design;
3. Exercise bar support arm;
4. Hemispherical dome structure;
5. Bridge structure with nondesign domain;

6. Single short corbel;
7. Double-sided beam-to-column joint;
8. Metallic insert;
9. Electric mast.

These examples demonstrate how topology optimization can be used to:

1. Provide several designs without an initial concepts from the designer;
2. Structurally improve a design concept;
3. Modify a concept maintaining certain features required by the designer;
4. Generate internal and external features in a design.

6.2 EFFECT OF DIFFERENT LOAD ANGLE ON A MICHELL CANTILEVER

This example explores what is the effect of varying the angle of the applied load on both the emerging topology and the final nondimensional mass for the Michell cantilever [22] of Section 5.2, with the length ratio $\overline{L} = 1.82196$. The Growth method of Chapter 3, Discrete Method of Structural Optimization, was used to solve this problem using the included program Truss Topology Optimization (TTO). Thirteen different angles in the range of between $\alpha = 0$ and 90 degrees were studied. The settings used by the program for all of the studied angles are given in Table 6.1 (Fig. 6.1).

TABLE 6.1 TTO Settings for All Models of the Michell Cantilever With Different Load Angle

L/H	σ_c/σ_t	Load	Angle	Iter.	Grid	Symmetry
1.82196	1	1	0° to −90°	1 to 180	50	No

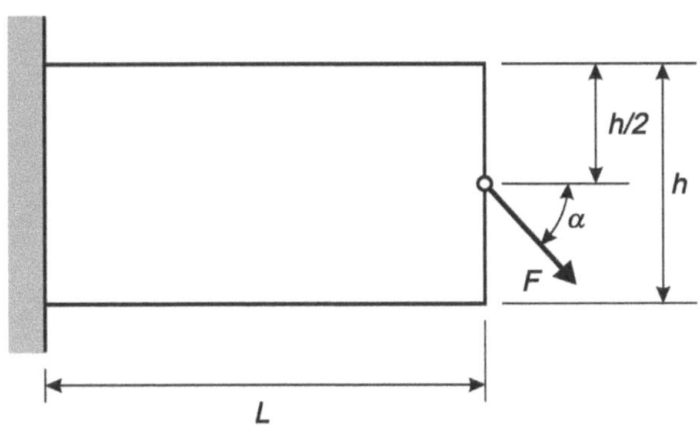

FIGURE 6.1 Design domain for the Michell cantilever with different load angle.

The optimal topologies for the different angles (α) are shown in Fig. 6.2, where the tensile bars are in red (black in print versions), the compressive bars in blue (dark grey in print versions) and any bar in green (light grey in print versions) represents a bar that has no cross-sectional area but is required to keep the structure statically determinate. The resulting increase in the nondimensional mass with increase in the angle of the applied force is given in Fig. 6.3.

As the angle between the force and the horizontal axis changes in the range of $0° \leq \alpha \leq 15.346°$; the only structure formed is a single beam aligned with the applied force. The reason for this is that a direct straight line exists between the left hand supporting wall and the line of action of the force. In the range of angles between $15.346° < \alpha < 37.55°$ this direct straight line no longer exists, and therefore the structure has to bend the applied force to an appropriate angle to connect it to the supporting side of the wall. This manifests itself as a fan like region emerging from the top left corner and consisting of straight radially positioned bars (Fig. 6.2C).

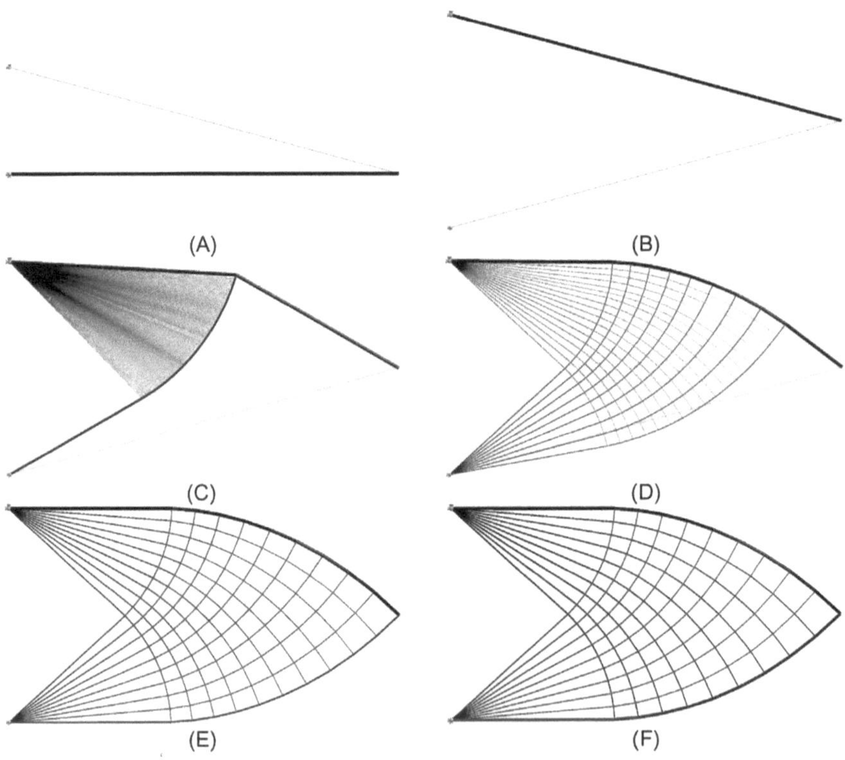

FIGURE 6.2 Emerging forms for the design of the Michell cantilever at different load angles: (A) 0°; (B) 15.346°; (C) 30°; (D) 37.55°; (E) 60°; and (F) 90°.

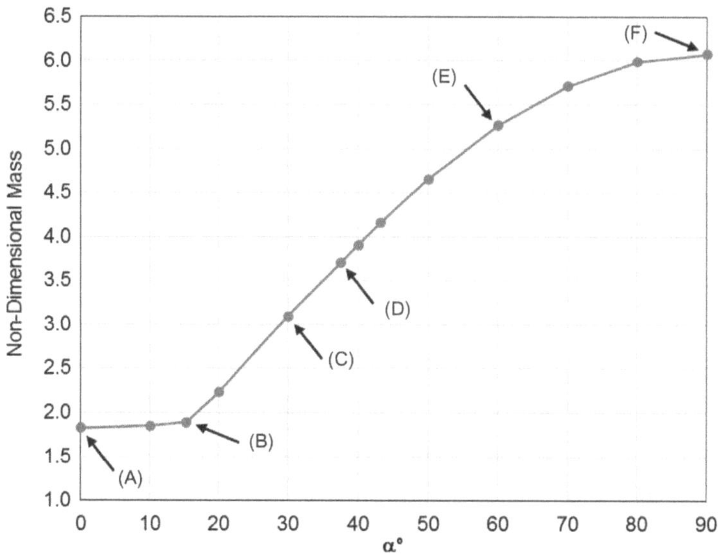

FIGURE 6.3 Variation of the nondimensional mass with respect to the angle of the applied force for the Michell cantilever.

For angles greater than 37.55 degrees; the emerging forms consist of two fan like regions of straight radially positioned bars from the two ends of the supporting location together with one large Hencky net region between the two fan regions and the point of application of the load (Fig. 6.2E and F). However between the angles of $37.55° < \alpha < 43.25°$, there is an extra bar connecting the applied force to the apex of the Hencky net region (Fig. 6.2D).

6.3 TAP OR FAUCET DESIGN

In this example, a designer has the task of developing a new form of a tap or faucet. The design can contain from between one to as many as eight arms, each of which can have any form. A sketch of the design space is shown in Fig. 6.4, with questions from the designer on the possible number and forms for the arms of the tap.

The 24 designs of Fig. 6.5 were generated using topology optimization, which always attempts to generate a form that is structurally sound and valid. The designs are presented such that the rows show the increasing number of arms (top to bottom) and the columns (left to right) show designs with increasing material volume. It can be seen in Fig. 6.5 that as less material is used for the design, the emerging forms become skeletal. All emerging forms however have biological characteristics, with for example the tap with five arms which resembles a star fish. Another emerging characteristic is that for

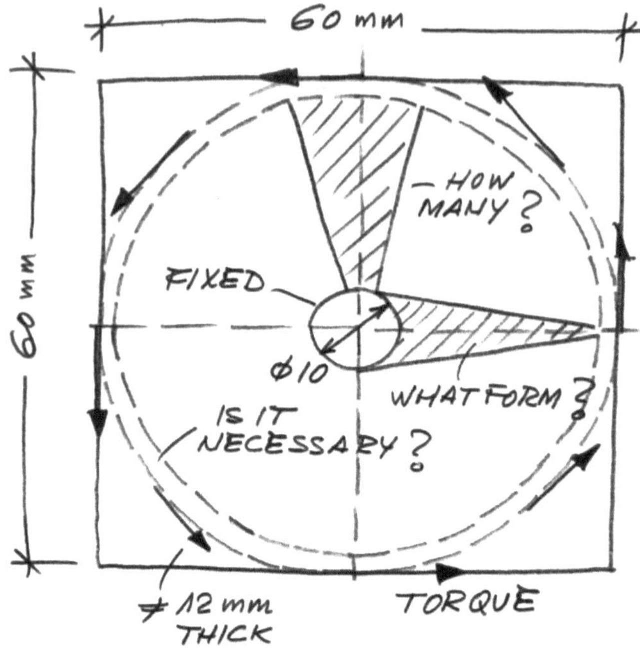

FIGURE 6.4 Sketch showing the design space and questions by the designer.

taps with more than one arm, the form of each arm resembles that of the single arm tap. Also, the internal features of each design have the form of the single arm design, which therefore resembles a fractal geometry. A further observation is that as the number of arms increase beyond five, the designs are very clean without intricate internal features.

The designs of Fig. 6.5 are all in two dimensions. In order to see how internal features are generated, a three dimensional optimization was carried out for the case with one arm present. The emerging forms are presented in Fig. 6.6. It can clearly be seen how an internal cavity forms and as designs are generated with less material, small skeletal structures emerge. The final design presented for the forms with least material is mainly skeletal in form with biological characteristics.

6.4 EXERCISE BAR SUPPORT ARM

This example illustrates the ability of topology optimization to modify a form conceived by a designer and maintain features or restrictions imposed on the design. The design consists of the support arm of an exercise bar. Fig. 6.7 shows that the initial design contains four lightening holes is trapezoidal in form and has a specific maximum size.

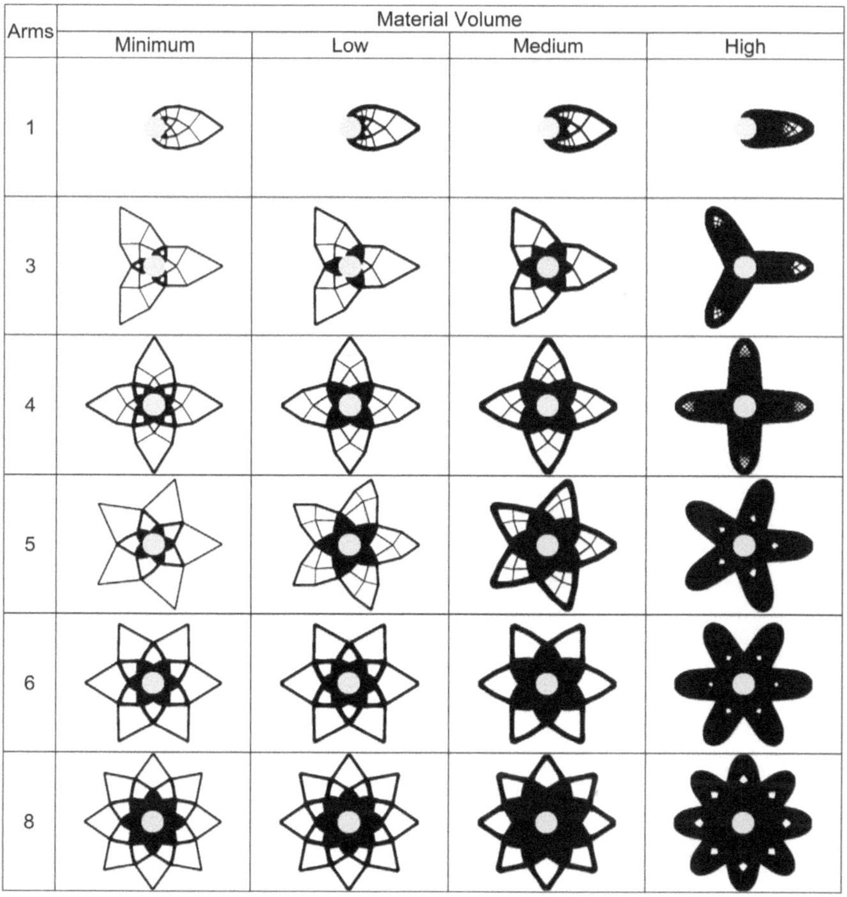

FIGURE 6.5 Emerging forms for the design of a tap or faucet.

Three topology optimizations were carried out, consisting of three different cases, to produce the results of Fig. 6.8. In case 1, the position and shape of the holes are unknown; in case 2 the position and size of the cavities are fixed; and in case 3 the designer had no initial suggestion for the design but had a restriction on the maximum space the design could occupy. The designs along each column for the three cases are presented with approximately the same material volume in decreasing material volume for ease of comparison. For case 1, topology optimization tries to produce the same designs as for the unrestricted case 3, but due to the trapezoidal design domain, the form is limited to these bounds. The designs for case 2 are interesting as the internal features are at the periphery of the boundary of where the cavities have been specified, but equally it is clear to see that by having specified their explicit location, they do influence the merging form. By

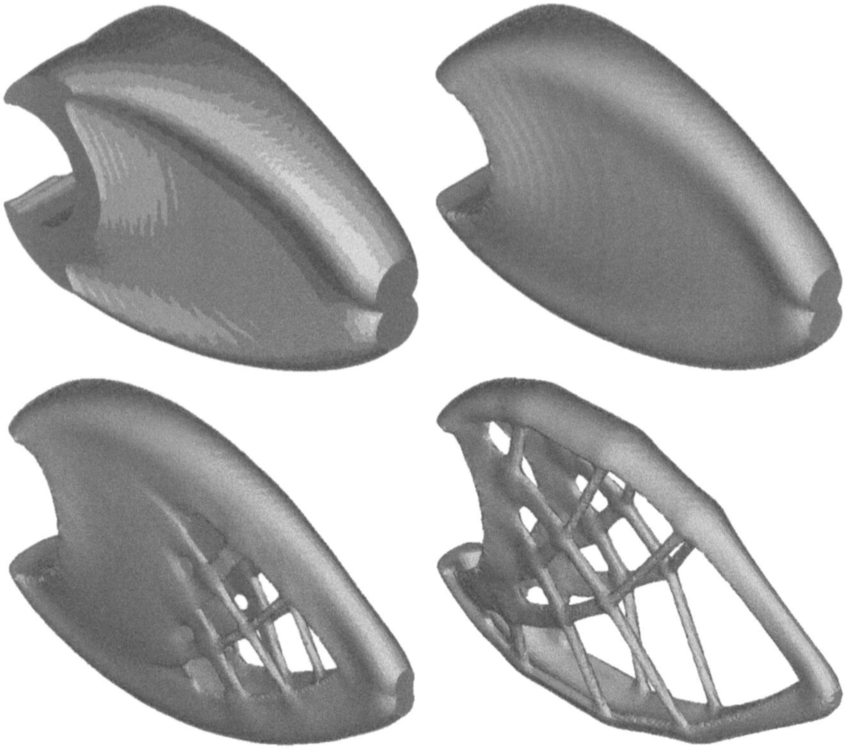

FIGURE 6.6 Isometric views of the emerging internal features for the design of a tap or faucet.

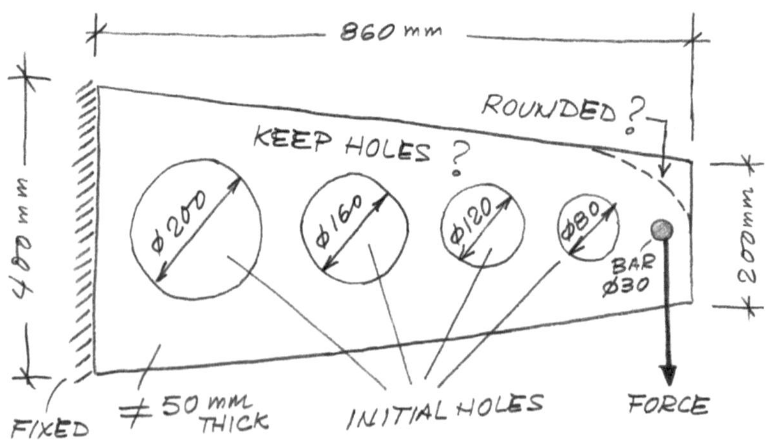

FIGURE 6.7 Sketch showing the design space and questions by the designer.

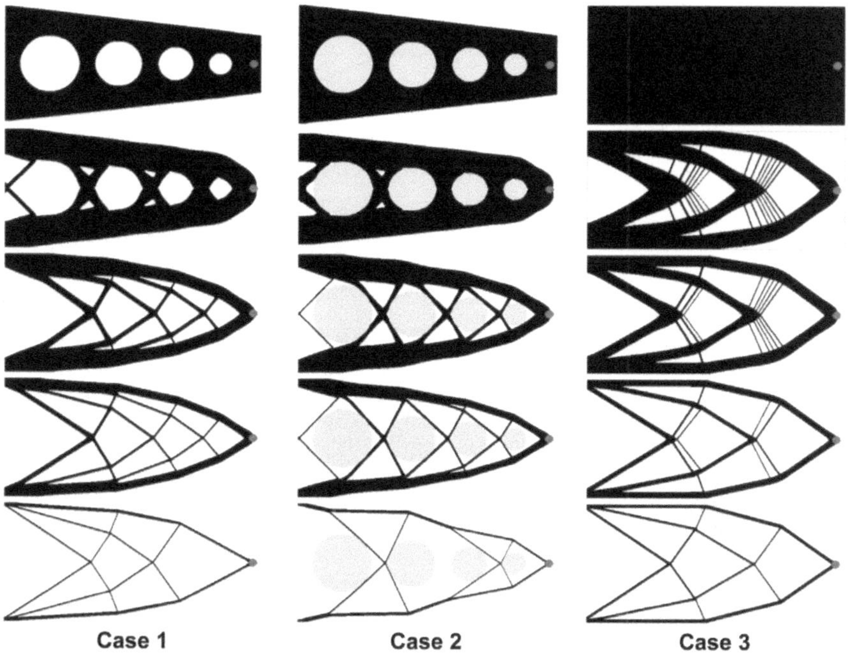

FIGURE 6.8 Emerging forms for the design of an exercise bar support arm.

specifying cavities at the beginning, the emerging designs are cleaner with less internal features than for the unrestricted case 3.

6.5 HEMISPHERICAL DOME STRUCTURE

This example illustrates the ability of topology optimization to aid a structural engineer in determining the location of the stiffeners which support a hemispherical dome. The design domain consists of a hemispherical shell with a diameter of 43.4 m and thickness $t = 2$ m, and with an opening at the top with a diameter of 9.1 m (Fig. 6.9).

The only loading is the self-weight of the initial design domain, with the only boundary condition consisting of the bottom circumference prevented from vertical displacements only. The material used for the stiffener was steel with an elastic modulus of 210 GPa, Poisson's ratio of 0.3 and a density of 7850 kg/m^3. The resulting topology can be seen in Fig. 6.10. The characteristics of the emergent structure are:

1. The generated design is in good agreement with the original solution to this problem which is the Pantheon in Rome [23]. However, the number of vertical and horizontal ribs (16 and 3, respectively) is significantly lower than in the Pantheon.

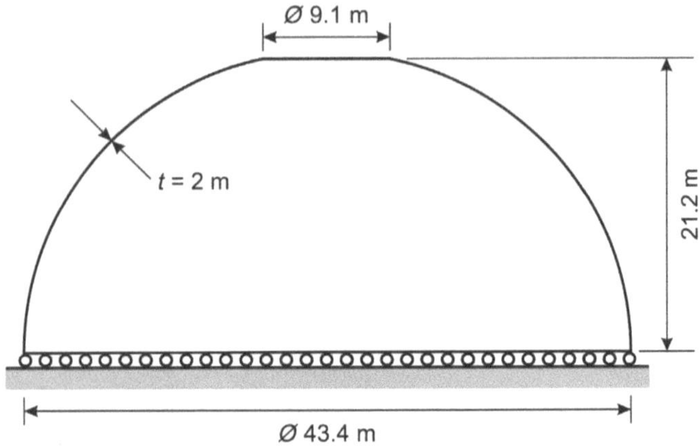

FIGURE 6.9 Dimensions of the hemispherical shell design domain.

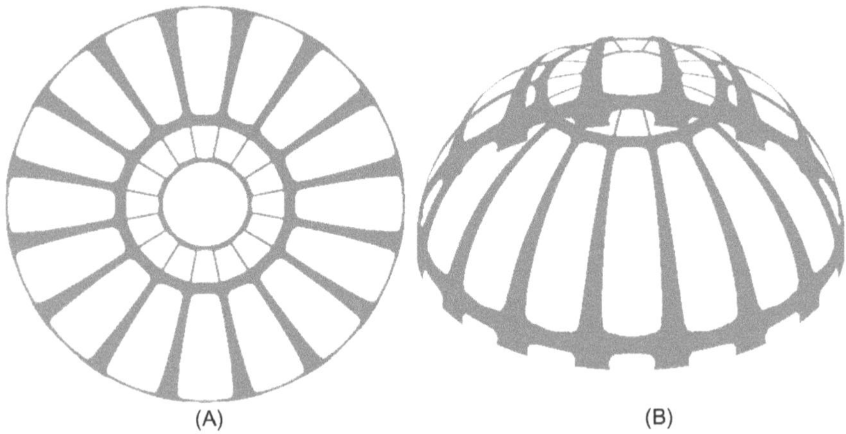

FIGURE 6.10 Hemispherical shell stiffener design: (A) top; (B) isometric views.

2. The horizontal rib located around the base (tension ring) is considerably larger than the others since the horizontal component (outward thrust) near the dome base is highest.

6.6 BRIDGE STRUCTURE WITH NONDESIGN DOMAIN

This example illustrates the ability of topology optimization to aid a structural engineer in determining the best layout of a bridge structure subject to a uniformly distributed load. The design domain consists of a rectangular region 180 m in length, 60 m high and 0.3 m thick (Fig. 6.11). The uniformly distributed load has a magnitude of 250 kN/m and acts in the middle of the section. The bridge deck is represented by a nondesign domain region 4 m deep below the loading. The continuum design was discretized using 270×90 elements.

The final design of the bridge which emerges for a final volume fraction of 20% is shown in Fig. 6.12. It can be seen that the section below the bridge deck was removed by the optimization process since it does not contribute to the structural stiffness. The optimal topology has the form of the well-known tie-arch bridge that is common in bridge engineering.

The maximum vertical displacement (D_{Max}) in mm; maximum (σ_{vM}^{max}), mean (σ_{vM}^{mean}), and minimum (σ_{vM}^{min}) von Mises stresses in MPa together with the nominal nondimensional mass for both the initial design domain and the final optimal form are given in Table 6.2. The nondimensional mass was

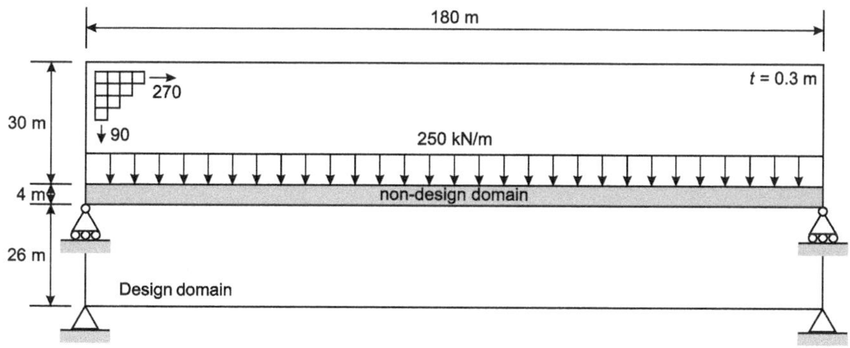

FIGURE 6.11 Bridge structure subjected to a uniformly distributed load.

FIGURE 6.12 Bridge structure design for final volume fraction of 20%.

TABLE 6.2 Bridge Structure Results for the Initial and Final Volume Fractions

(V/V_0)	(D_{Max})	(σ_{vM}^{max})	(σ_{vM}^{mean})	(σ_{vM}^{min})	$\overline{M}$
1	2.428	31.42	2.05	0.79	1.53067
0.2	10.680	60.27	10.29	6.6	1.53664

calculated using Eq. (5.2) with the total applied force $F = 45$ MN and using the semispan as the nominal length $L_{nom} = 90$ m, which is the same dimension used for the Michell beams of Sections 5.5 and 5.6.

6.7 SINGLE SHORT CORBEL

This example illustrates the ability of topology optimization to aid a structural engineer to determine the best position and material distribution for steel reinforcement of a concrete structure. A corbel is a short cantilever beam attached to a column which carries a point load a distance away from the column centreline. If the distance between the load application point and the column face is less than the height of the corbel, it is called a short corbel. If only the short corbel was modeled, its boundary condition would significant affect the load path through it. For this reason it is common practice [24] to model the short corbel with a section of the column on either side of it (Fig. 6.13).

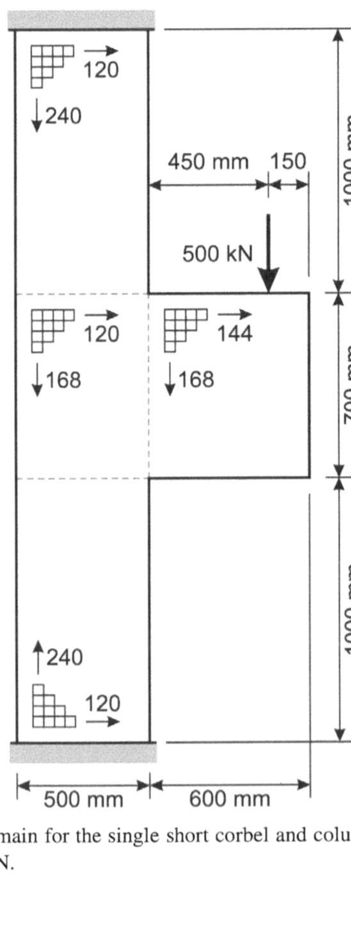

FIGURE 6.13 Design domain for the single short corbel and column subjected to a downwards concentrated load of 500 kN.

The column is fixed at both ends, and the complete structure has a thickness of 300 mm. A downwards force of 500 kN is applied on the short corbel at a distance of 450 mm from the column face. The structure was divided using square finite elements of length 4.17 mm, using a total of 101,952 elements. The elasticity modulus and density for the compressive elements (concrete) were $E_C = 20$ GPa and $\rho_C = 2000$ kg/m^3 and for the tensile elements (steel) were $E_T = 210$ GPa and $\rho_T = 7850$ kg/m^3. Two different problems were investigated: (A) Considering the same material properties for the tension and compressive regions and (B) Considering different material properties for the tension and compressive regions and using Eq. (4.7) to give the stress ratio of $k_1 = 6.42$.

The resulting forms of the designs are given in Fig. 6.14, where the tensile regions are in red (black in print versions) and the compressive regions in blue (grey in print versions). Fig. 6.14(A) shows a classic strut-and-tie model which is in excellent agreement with the analytical result of Schlaich et al. [24] and the numerical results of Liang et al. [25]; Kwak and Noh [26], and Bruggi [27]. Fig. 6.14B shown the T&C model, so named as different material properties are used for the tensile and compressive regions [28].

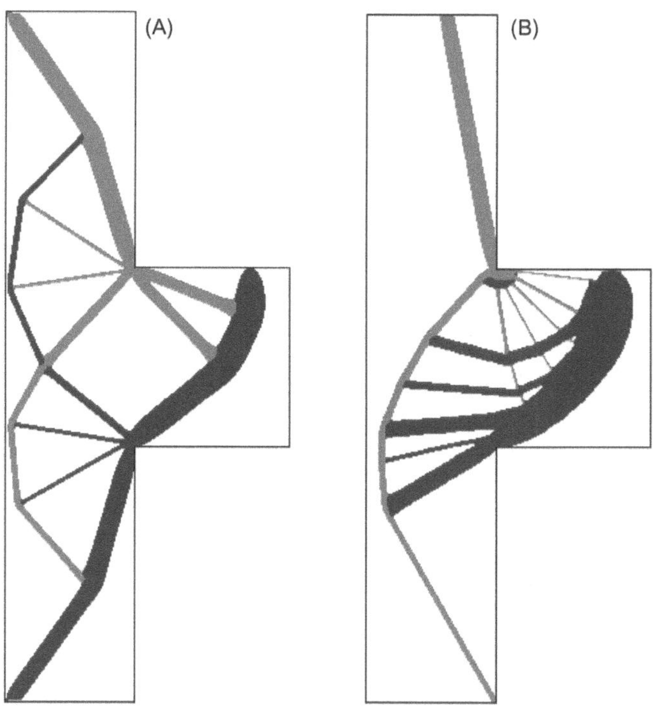

FIGURE 6.14 Strut-and-tie models for the single short corbel for a volume fraction of 20%. (A) Classic model ($k_1 = 1.0$), (B) T&C model ($k_1 = 6.42$).

This resulting form is considerably different than the classic model and clearly shows a structural engineer where to place the reinforcing steel material to strengthen and reinforce the concrete structure.

6.8 DOUBLE-SIDED BEAM-TO-COLUMN JOINT

This example illustrates the ability of topology optimization to aid a structural engineer to determine the best position and material distribution for the reinforcement of concrete structures, which mainly consist of beam-and-column frameworks. One type of joint in these frameworks is the double-sided beam-to-column (also named T-joint), which consists of a continuous beam supported on a column (Fig. 6.15). The bottom of the column is fixed and the complete structure has a thickness of 300 mm. The loading consists of two bending moments applied at the free ends of the beam, each simulated by two opposing forces 400 mm apart. The structure was divided using

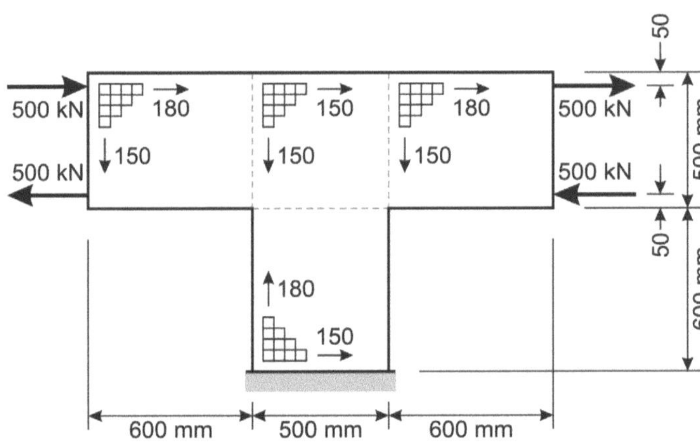

FIGURE 6.15 Design domain of the double-sided beam-to-column joint subjected to bending moments simulated by two opposing 500 kN forces.

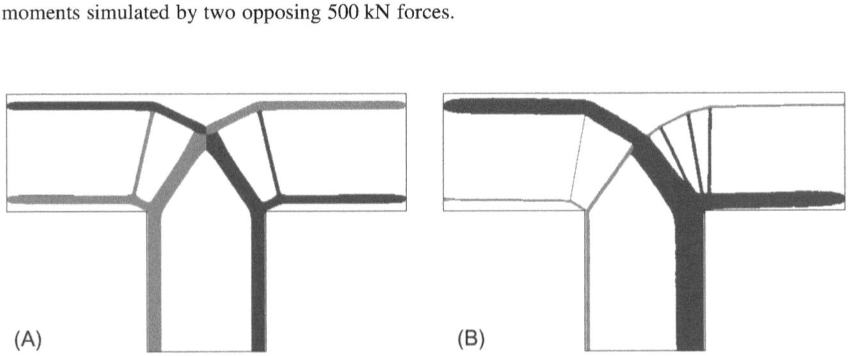

FIGURE 6.16 Strut-and-tie models for the double-sided beam-to-column joint for a volume fraction of 22%: (A) classic model ($k_1 = 1.0$), (B) T&C model ($k_1 = 6.42$).

square finite elements of length 3.33 mm, using a total of 103,500 elements. The material properties are the same as those of the example of Section 6.7.

The resulting forms of the designs for a volume fraction of 22% are given in Fig. 6.16, where the tensile regions are in red (black in print versions) and the compressive regions in blue (grey in print versions). The classic symmetrical strut-and-tie topology model is given in Fig. 6.16A. The T&C model which shows that the topology should not be symmetrical about the vertical-axis is shown in Fig. 6.16B, the reason for the lack of symmetry is due to the redistribution of the tensile and compressive regions.

6.9 METALLIC INSERT

This example illustrates the ability of topology optimization to aid an engineer to design the best shape for a metal insert inside of a composite material structure. The dimensions, loading and support conditions for this problem are shown in Fig. 6.17. Due to symmetry, the optimization of the form was made using only the top-half of the domain using a regular rectangle mesh [29]. To allow for available fasteners, the nearest region around the boundary of the hole was defined as nondesign domain. This ensured that the internal diameter remained unchanged. The bolt hole loading was represented by a linearly varying pressure distribution (Fig. 6.17B). The metal insert was aluminum with properties $E_{\mathrm{Al}} = 71$ GPa and $\nu_{\mathrm{Al}} = 0.3$. The surrounding structural domain used the properties proposed by Rispler and Steven [29] where $E_{\mathrm{domain}} = 52.2$ GPa and $\nu_{\mathrm{domain}} = 0.33$.

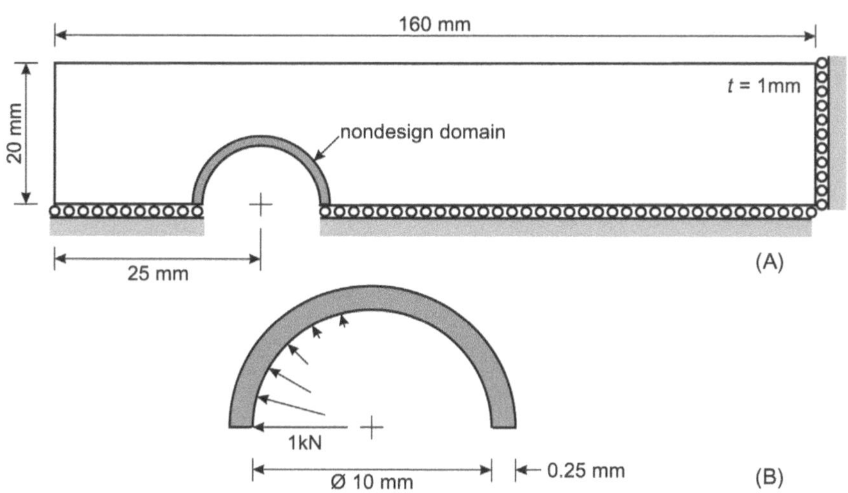

FIGURE 6.17 Design domain for the metallic insert: (A) dimensions and support conditions; (B) detailed view of the applied linearly varying load.

The resulting designed forms for two ratios between the aluminum insert area (A_{Al}) and the area of the hole (A_{Hole}) are given in Fig. 6.18: $A_{Al}/A_{Hole} = 1{:}1$ and 3:1. The shape of the aluminum metal insert is in light blue (grey in print versions), the nondesign domain is in orange (black in pring versions) and the composite material is not shown. The obtained shape of the inserts agree with the solutions obtained by Rispler and Steven [29].

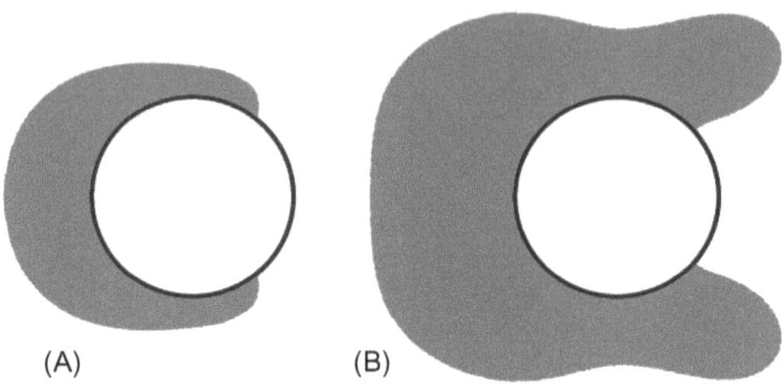

FIGURE 6.18 Final form for the aluminum insert (in light blue, grey in print versions) for (A_{Al}/A_{Hole}): (A) 1:1 and (B) 3:1, with the nondesign domain shown in orange.

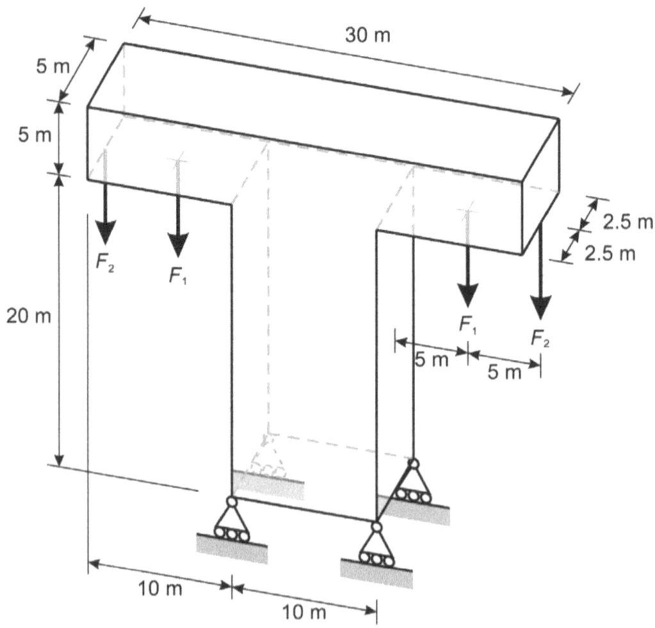

FIGURE 6.19 Design domain for the electric mast.

6.10 ELECTRIC MAST

This example illustrates the ability of topology optimization to aid an engineer to design the general form of an electric mast. The design domain is defined as the T-shaped box of Fig. 6.19. Two symmetrical vertical loads F_1 and F_2, both of magnitude 10 kN, were applied in the centreline of the lower edges of the horizontal part of the T-section and represent the loads exerted by the wires on the mast. These loads were applied either as single load cases or as multiple load cases. Simply supported boundary conditions were applied at the four corners of the base.

The resulting forms of the designs are shown in Fig. 6.20. The case with only force F_1 acting is shown in Fig. 6.20A. The case with only force F_2 acting is shown in Fig. 6.20B. The case with both forces F_1 and F_2 acting as a single load case is shown in Fig. 6.20C. The multiple load case where forces F_1 and F_2 act independently of each other is shown in Fig. 6.20D. These

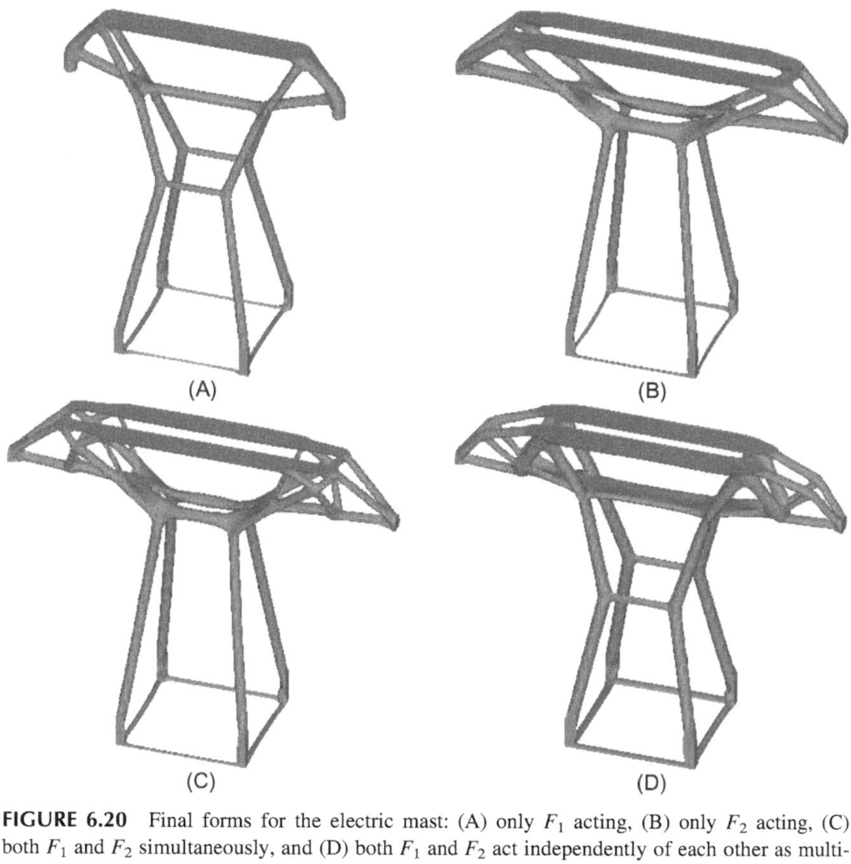

FIGURE 6.20 Final forms for the electric mast: (A) only F_1 acting, (B) only F_2 acting, (C) both F_1 and F_2 simultaneously, and (D) both F_1 and F_2 act independently of each other as multiple load cases.

designed forms are all truss-like structures where the number of bars or truss elements are dependent, to an extent, on the mesh density and size of the fixed grid domain.

REFERENCES

[1] ANSYS, Ansys 17.2 [Software]. Available from: <http://www.ansys.com/>, 2016 (accessed 23.12.16).
[2] SIMULIA, Abaqus 6.14 [Software]. Available from: <http://www.intrinsys.com/software/simulia/abaqus>, 2016 (accessed 23.12.16).
[3] Dynamic Design Solutions, FEMtools 3.8, [Software]. Available from: <http://www.fem-tools.com/>, 2016 (accessed 23.12.16).
[4] MSC Software, MSC Nastran 2016 [Software]. Available from: <http://www.mscsoftware.com/product/msc-nastran>, 2016 (accessed 23.12.16).
[5] Altair, HyperWorks OptiStruct [Software], Available from: <http://www.altairhyperworks.com/product/OptiStruct>, 2016 (accessed 23.12.16).
[6] Creative Engineering Services, CATOPO [Software]. Available from: <www.catopo.com>, 2016 (accessed 23.12.16).
[7] TopOpt Research Group [Online]. Available from: <http://www.topopt.dtu.dk/>, 2016 (accessed 23.12.16).
[8] CalculiX [Online]. Available from: <http://www.calculix.de/>, 2016 (accessed 23.12.16).
[9] R. Oxman, Digital architecture as a challenge for design pedagogy: theory, knowledge, models and medium, Des. Stud. 29 (2008) 99–120.
[10] R. Oxman, Theory and design in the first digital age, Des. Stud. 27 (2006) 229–265.
[11] B. Kolarevic, Architecture in the Digital Age, Spon Press, New York, 2003.
[12] B. Kolarevic, Designing and manufacturing architecture in the digital age, in: Proceedings of the 19th Education for Computer Aided Architectural Design in Europe (eCAADe19), Session 05 Design Process, pp. 117–123, 2001.
[13] H. Kim, O.M. Querin, G.P. Steven, On the development of structural optimisation and its relevance in engineering design, Des. Stud. 23 (2002) 85–102.
[14] O. Sigmund, M.P. Bendsøe, Topology optimization – from airplanes to nanooptics, in: K. Stubkjær, T. Kortenbach (Eds.), Part of: BRIDGING From Technology to Society, Technical University of Denmark, Lyngby, Denmark, 2004, pp. 40–51. (ISBN: 87-990378-0-7).
[15] L. Krog, A. Tucker, G. Rollema, Application of Topology, Sizing and Shape Optimization Methods to Optimal Design of Aircraft Components, Airbus UK Ltd., Altair Engineering Ltd., 2002.
[16] Y.M. Xie, P. Felicetti, J.W. Tang, M.C. Burry, Form finding for complex structures using evolutionary structural optimization method, Des. Stud. 26 (2005) 55–72.
[17] F. Poncelet, C. Fleury, A. Remouchamps, S.Grihon, TOPOL: a topological optimization tool for industrial design, in: Proceedings of the 6th World Congress of Structural and Multidisciplinary Optimization (WCSMO6), Rio de Janeiro, Brazil, Paper 5141, 2005.
[18] S.L. Lee, D.C. Lee, J.I. Lee, C.S. Hand, K. Hedricke, Integrated process for structural–topological configuration design of weight-reduced vehicle components, Finite Elem. Anal. Des. 43 (2007) 620–629.
[19] W. Wu, D.Z. Yang, Y.Y. Huang, M. Qi, W.Q. Wang, Topology optimization of a novel stent platform with drug reservoirs, Med. Eng. Phys. 30 (2008) 1177–1185.

[20] L. Frattari, R. Vadori, R. D'Aria, G. Leoni, Topology optimization in architecture: May it be a design tool? in: Proceedings of the 8th World Congress on Structural and Multidisciplinary Optimization (WCSMO8), Lisbon, Portugal, Paper 1502, 2009.

[21] L. Frattari, R. Vadori, R. D'Aria, L. Pietroni, An Eco-Design-Oriented Multidisciplinary Approach in Industrial Design, in: Proceedings of the 8th World Congress on Structural and Multidisciplinary Optimization (WCSMO8), Lisbon, Portugal, Paper 1501, 2009.

[22] G.I.N. Rozvany, Exact analytical solutions for some popular benchmark problems in topology optimization, Struct. Optim. 15 (1998) 42–48.

[23] Ancient History Encyclopedia [Online]. Available from: <http://www.ancient.eu/Pantheon/>, 2016 (accessed 23.12.16).

[24] J. Schlaich, K. Schäfer, M. Jennewein, Toward a consistent design of structural concrete, PCI J. 32 (3) (1987) 75–150.

[25] Q.Q. Liang, Y.M. Xie, G.P. Steven, Topology optimization of strut-and-tie models in reinforced concrete structures using an evolutionary procedure, ACI Struct. J. 97 (2) (2000) 322–330.

[26] H.G. Kwak, S.H. Noh, Determination of strut-and-tie models using evolutionary structural optimization, Eng. Struct. 28 (2006) 1440–1449.

[27] M. Bruggi, Generating strut-and-tie patterns for reinforced concrete structures using topology optimization, Comput. Struct. 87 (2009) 1483–1495.

[28] M. Victoria, O.M. Querin, P. Martí, Generation of strut-and-tie models by topology design using different material properties in tension and compression, Struct. Multidisc. Optim. 44 (2011) 247–258. Available from: http://dx.doi.org/10.1007/s00158-011-0633-z.

[29] A.R. Rispler, G.P. Steven, Shape optimisation of metallic inserts in composite bolted joints, in: Proceedings of the PICAST 2 – AAC 6, Melbourne, 1995.

Chapter 7

User Guides for Enclosed Software*

7.1 INTRODUCTION

This chapter provides the user guides to the different programs which implement the topology optimization methods presented in Chapters 2, 3 and 4. The program that implements the Growth method for the size, topology, and geometry optimization of truss structures is called TTO and its user guide is presented in Section 7.2, with examples of its use for different problems explained in Section 7.3. The MATLAB [1] *"SERA.m"* file that implements the Sequential Element Rejection and Admission (SERA) method is presented in Section 7.4. This program is then manipulated in Section 7.5 to show how to modify it for different topology optimization problems. The software that implements the Isolines/Isosurfaces Topology Design (ITD) method is called liteITD and is presented in Section 7.6. Examples of how to use liteITD are presented in Sections 7.7 and 7.8. With Section 7.9 giving consistent sets of units which can be used in the liteITD program.

7.2 TRUSS TOPOLOGY OPTIMIZATION (TTO) PROGRAM

The TTO program incorporates into a single executable program the method presented in Chapter 2, Growth Method for the Size, Topology, and Geometry Optimization of Truss Structures. This section presents all aspects of the program; its installation; what its Graphical User Interface (GUI) looks like and what all of its features are, thus providing a basic user guide for TTO.

7.2.1 System Requirements and Installation of TTO

The TTO program is a 32-bit application for the Windows operating system. The version included with this book was developed in 2010 and can be used

* Additional author for this chapter is Dr. Pedro Jesús Martínez Castejón.

Topology Design Methods for Structural Optimization.
DOI: http://dx.doi.org/10.1016/B978-0-08-100916-1.00007-6

to generate all of the examples presented in this book. To install the TTO program, please follow these steps:

1. Using Windows Explorer, create a directory called "\TTO" in your computer.
2. Please access the companion website for this book: https://www.elsevier.com/books-and-journals/book-companion/9780081009161 and please enter the "TTO software" link."
3. From the "TTO software" link download the zipped file "TTOSoftware.zip" into the TTO directory created in your computer in step (1), and unzip the file in the same directory. The list of all files is given in Table 7.1.
4. The TTO program corresponds with the file named "*Tto.exe*," Fig. 7.1. Please note that the *"Tto.dll"* file must always be located in the same directory as the *"Tto.exe"* file for the program to work.
5. Create a shortcut of the TTO program by right clicking with the mouse button on the "*Tto.exe*" icon, then select *Send to -> Desktop (create shortcut)*.
6. Installation is now complete.
7. To start TTO, double-click on the shortcut of the icon on the Desktop.

TABLE 7.1 List of All Files in the "TTOSoftware.zip" file from this Book Companion Website

Name	Size
MbbBeam_2.40196.dat	2 kB
MbbBeam_5.49846.dat	1 kB
MichellBeam_FixedSupports.dat	1 kB
MichellBeam_RollerSupports.dat	1 kB
MichellCantilever_0.5.dat	1 kB
MichellCantilever_1.82196.dat	1 kB
MichellCantilever_3.35889.dat	1 kB
MichellCantilever_CircularBoundary_Lr10.dat	1 kB
MichellCantilever_CircularBoundary_Lr15.dat	1 kB
PragerCantilever_108Iterations.dat	1 kB
PragerCantilever_12Iterations.dat	1 kB
Tto.dll	1608 kB
Tto.exe	6075 kB

FIGURE 7.1 TTo program.

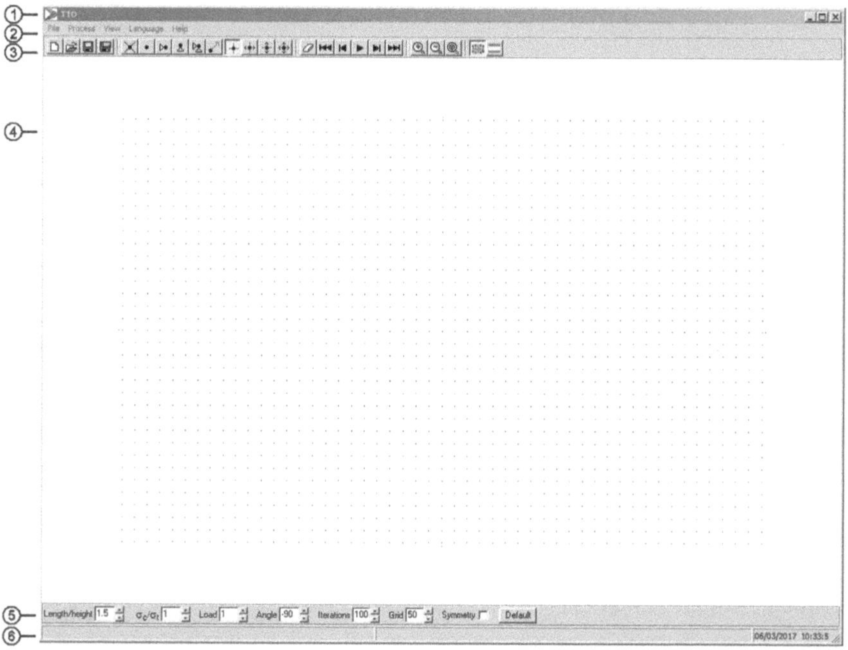

FIGURE 7.2 The TTo Graphical User Interface (GUI).

7.2.2 Overview of the TTO Graphical User Interface

When TTO is started, the GUI of Fig. 7.2 appears. This is similar to most Windows applications as it has most of the usual features. From top to bottom, the GUI consists of six sections: (1) Title bar; (2) Menu bar; (3) Button bar; (4) Display area; (5) Program options bar; and (6) Status bar.

7.2.2.1 Title Bar

The Title bar (part 1 in Fig. 7.2) displays the name of the application "*TTO*." On the right hand side, it also has the standard Windows buttons to minimize, maximize, and close the application.

7.2.2.2 Menu Bar

The Menu bar (part 2 in Fig. 7.2) has five menus: (1) File; (2) Process; (3) View; (4) Language; and (5) Help. These menus are explained in the following sections.

When *"File"* is selected, the drop-down menu list of Fig. 7.3 is displayed. There are five commands in the File menu: (1) New; (2) Open; (3) Save as; (4) Save screen as; and (5) Exit. The actions carried out by each menu command are described next.

1. *New:* This command sets all of the program options to their default value in readiness for a new optimization problem.
2. *Open:* This command opens an existing data file which contains a structural model to be optimized.
3. *Save as:* This command saves the current program options as well as the structural domain, supports and loads to a data file. *Note that although a file name needs to be used so that the program can save all relevant information, there is a known bug in the way that TTO names the file. Although a file name needs to be typed here, the program instead saves all of the relevant data to a file which consists of a number with no file extension. The file should then be renamed using Windows Explorer and given the extension ".dat". This file can then be opened normally by TTO and will contain all of the model data.*
4. *Save screen as:* This command saves an instance of everything which is in the display area into an image bitmap file with the extension "*.bmp*". The filename defaults to the current structure filename specified in the *Save as* command of step (3). Alternatively a new name can be specified.
5. *Exit:* This command will exit the program.

When *"Process"* is selected, the drop-down menu list of Fig. 7.4 is displayed. There are six commands in the Process menu: (1) Clear; (2) First;

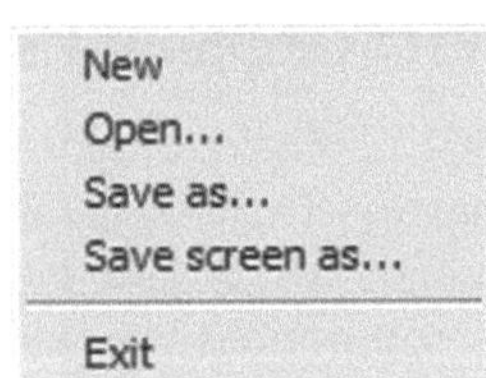

FIGURE 7.3 File drop-down menu list.

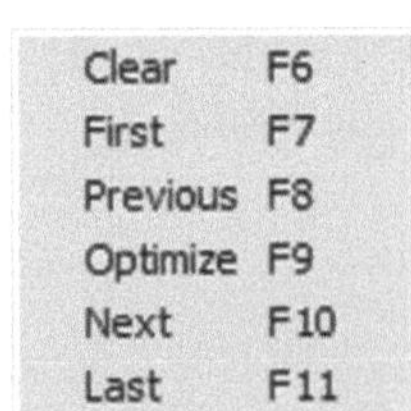

FIGURE 7.4 Process drop-down menu list.

(3) Previous; (4) Optimize; (5) Next; and (6) Last. The actions carried out by each menu command are described next.

1. *Clear:* This command clears an optimization run and resets the iteration counter to 1.
2. *First:* This command displays the topology obtained in the first iteration of the optimization process.
3. *Previous:* This command displays the topology from the iteration number prior to the current one.
4. *Optimize:* This command launches the optimization process of the structure, from either iteration 1, or if the optimization process had been paused, from the last iteration number reached.
5. *Next:* This command displays the topology for the next iteration number. If the optimization is being run one iteration at a time, then TTO will carry out the next optimization iteration and at the end, display the next topology. If, however, the user had paused the optimization process to investigate previous optimal topologies, this command will display the next in the list of stored optimized topologies.
6. *Last:* This command displays the topology obtained in the last iteration of the optimization process.

When *"View"* is selected, the drop-down menu list of Fig. 7.5 is displayed. There are seven commands in the View menu: (1) Grid; (2) Joints; (3) Supports; (4) Loads; (5) Zoon in; (6) Zoom out; and (7) Zoom all. The actions carried out by each menu command are described next.

1. *Grid:* If selected (denoted by the ✓), the background grid will be shown in the display area.
2. *Joints:* If selected (denoted by the ✓), the joints connecting each bar will be shown in the display area.
3. *Supports:* If selected (denoted by the ✓), the ground supports will be shown in the display area.
4. *Loads:* If selected (denoted by the ✓), the applied loads will be shown in the display area.

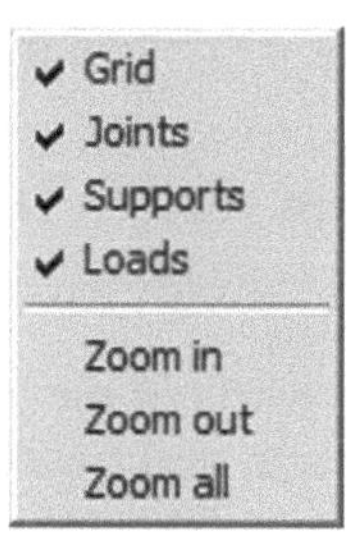

FIGURE 7.5 View drop-down menu list.

FIGURE 7.6 Language drop-down menu list.

FIGURE 7.7 Help drop-down menu list.

5. *Zoom in:* This command zooms in by a factor of 1.05, thus increasing the size of the display area by 5%.
6. *Zoom out:* This command zooms out by a factor of 0.95, thus reducing the size of the display area by 5%.
7. *Zoom all:* This command resets the display area to the default setting, where the working design domain (Section 7.2.2.4) is approximately 80% of the display area.

When *"Language"* is selected, the drop-down menu list of Fig. 7.6 is displayed. There are two commands in the Language menu: (1) English and (2) Spanish. The actions carried out by each menu command are described next.

1. *English:* If selected (denoted by the ✓), the language used for all commands and displays in the program is set to English (*Default language setting*).
2. *Spanish:* If selected (denoted by the ✓), the language used for all commands and displays in the program is set to Spanish.

When *"Help"* is selected, the drop-down menu list of Fig. 7.7 is displayed. There is only one commands in the Help menu: (1) About. The effect of selecting this item is explained below.

About: This command opens the window of Fig. 7.8 with information about the TTO program, its author, and contact details.

7.2.2.3 Button Bar

The Button bar (part 3 in Fig. 7.2) has 25 buttons and is divided into five sections. The five sections are called: (1) File; (2) Joint functions; (3) Process; (4) Zoom; and (5) Language. The following sections explain the actions carried out by these buttons.

The *"File"* button section consists of the four buttons of Fig. 7.9. These correspond with exactly the same four commands in the File menu of Section 7.2.2.2: (1) New; (2) Open; (3) Save as; and (4) Save screen. These buttons are explained next.

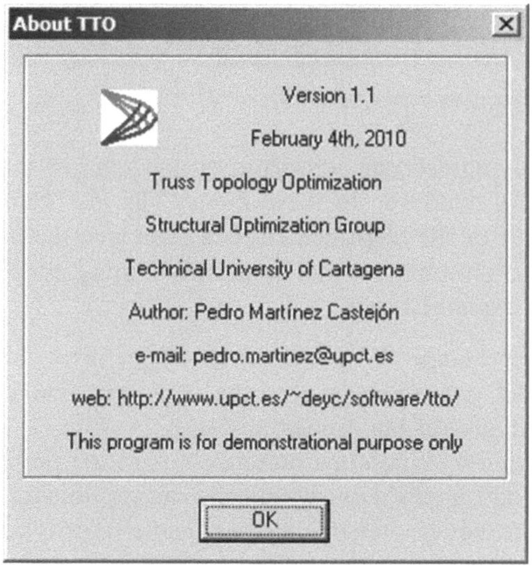

FIGURE 7.8 About information window.

FIGURE 7.9 File button menu list.

1. *New button:* Selecting this button sets all of the program options to their default value in readiness for a new optimization problem.
2. *Open button:* Selecting this button opens an existing data file which contains the structural model to be optimized
3. *Save as button:* Selecting this button saves the current program options as well as the structural domain, supports and loads to a data file. *Note that although a file name needs to be used so that the program can save all relevant information, there is a known bug in the way that TTO names the file. Although a file name needs to be typed here, the program instead saves all of the relevant data to a file which consists of a number with no file extension. The file should then be renamed using Windows Explorer and given the extension ".dat". This file can then be opened normally by TTO and will contain all of the model data.*
4. *Save screen as button:* Selecting this button saves an instance of everything which is in the display area into an image bitmap file with the extension *".bmp"*. The filename defaults to the current structure filename specified in the *Save as button* command of step (3). Alternatively a new name can be specified.

The *"Joint functions"* button section consists of the 10 buttons of Fig. 7.10, arranged into two different sections. The first six buttons control

FIGURE 7.10 Joint functions button menu list.

the action of applying/deleting joints/forces and control the structural displacement boundary condition of these joints. The last four buttons control the movement limit or the displacement constraints of the geometry optimization cycle of the program for any joint added using the first six buttons. These buttons are explained next.

1. *Delete a joint button:* When this button  is in the pressed down mode, the user can delete any of the user specified joints within the design domain grid of the display area.
2. *Add a joint button:* When this button  is in the pressed down mode, the user can add a new joint within the design domain grid of the display area. The newly added joint is capable of displacements in both the *X*- and *Y*- axes due to the applied load(s).
3. *Add a joint with fixed X-displacement button:* When this button  is in the pressed down mode, the user can add a new joint within the design domain grid of the display area. The newly added joint is constrained from moving in the *X*-axis but is capable of displacement in the *Y*-axis due to the applied load(s).
4. *Add a joint with fixed Y-displacement button:* When this button  is in the pressed down mode, the user can add a new joint within the design domain grid of the display area. The newly added joint is constrained from moving in the *Y*-axis but is capable of displacement in the *X*-axis due to the applied load(s).
5. *Add a joint with all fixed displacements button:* When this button  is in the pressed down mode, the user can add a new joint within the design domain grid of the display area. The newly added joint is constrained from all movement in both the *X*- and *Y*- axes due to the applied load(s).
6. *Add a joint with an applied load button:* When this button  is in the pressed down mode, the user can add a new joint within the design domain grid of the display area. The newly added joint contains an applied force with the magnitude specified in the *Load magnitude edit box* and with the direction specified in the *Load angle edit box* which are located in the Program options bar (Section 7.2.2.5).
7. *All fixed coordinates button:* When this button 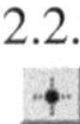 is in the pressed down mode (the default condition for all new nodes), every new joint added within the design domain grid of the display area (buttons 1–6 above) has the move limit for the geometry optimization set to zero in all directions (i.e., prevented from changing its position during the geometry optimization cycle).

8. *Free X coordinates button:* When this button is in the pressed down mode, every new joint added within the design domain grid of the display area (buttons 1–6 above) has the move limit for the geometry optimization set to zero in the *Y*-direction only (i.e., prevented from changing its position along the *Y*-direction but allowed to be repositioned along the *X*-direction during the geometry optimization cycle).
9. *Free Y-coordinate button:* When this button is in the pressed down mode, every new joint added within the design domain grid of the display area (buttons 1–6 above) has the move limit for the geometry optimization set to zero in the *X*-direction only (i.e., prevented from changing its position along the *X*-direction but allowed to be repositioned along the *Y*-direction during the geometry optimization cycle).
10. *All Free coordinates button:* When this button is in the pressed down mode, every new joint added within the design domain grid of the display area (buttons 1–6 above) has no move limit imposed for the geometry optimization (i.e., allowed to change its position along both the *X*- and *Y*-directions during the geometry optimization cycle).

The *"Process"* button section consists of the six buttons of Fig. 7.11. These correspond with exactly the same six commands in the Process menu of Section 7.2.2.2: (1) Clear; (2) First; (3) Previous; (4) Optimize; (5) Next; and (6) Last. These buttons are explained next.

1. *Clear button:* Selecting this button clears an optimization run and resets the iteration counter to 1.
2. *First button:* Selecting this button displays the topology obtained in the first iteration of the optimization process.
3. *Previous button:* Selecting this button displays the topology from the iteration number prior to the current one.
4. *Optimize button:* Selecting this button launches the optimization process of the structure, from either iteration 1, or if the optimization process had been paused, from the last iteration number reached. Whenever the optimization process is launched, this button changes to the pause button, which when pressed, pauses the optimization.
5. *Next button:* Selecting this button displays the topology for the next iteration number. If the optimization is being run one iteration at a time, then TTO will carry out the next optimization iteration and at the end, display the next topology. If, however, the user had paused the optimization process to investigate previous optimal topologies, this command will display the next in the list of stored optimized topologies.

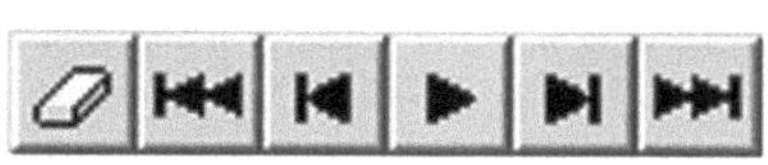

FIGURE 7.11 Process button menu list.

FIGURE 7.12 Zoom button menu list.

FIGURE 7.13 Language button menu list.

6. *Last button:* Selecting this button displays the topology obtained in the last iteration of the optimization process.

The "*Zoom*" button section consists of the three buttons of Fig. 7.12. These correspond with exactly the same last three commands from the View menu of Section 7.2.2.2: (1) Zoon in; (2) Zoom out; and (3) Zoom all. These buttons are explained next.

1. *Zoom in button:* Selecting this button zooms in by a factor of 1.05, thus increasing the size of the display area by 5%.
2. *Zoom out button:* Selecting this button zooms out by a factor of 0.95, thus reducing the size of the display area by 5%.
3. *Zoom all button:* Selecting this button resets the display area to the default setting, where the working design domain (Section 7.2.2.4) is approximately 80% of the display area.

The "*Language*" button section consists of the two buttons of Fig. 7.13. These correspond with exactly the same two commands as in the Language menu of Section 7.2.2.2: (1) English and (2) Spanish. These buttons are explained next.

1. *English button:* Selecting this button sets English as the language used for all commands and displays in the program (*Default language setting*).
2. *Spanish button:* Selecting this button sets Spanish as the language used for all commands and displays in the program.

7.2.2.4 Display Area Bar

The Display area (part 4 in Fig. 7.2) contains inside of it the working design domain grid with an aspect ratio specified in the *Design domain aspect ratio edit box* and where the number of grid points along the horizontal *X*-axis is specified in the *Grid edit box* which is also located in the Program options bar (Section 7.2.2.5). In order to aid the placement of joints and loads, there is an extra grid point next to the central grid point along each of the four sides. The working design domain grid occupies approximately 80% of the display area.

FIGURE 7.14 Program options bar items.

7.2.2.5 Program Options Bar

The Program options bar (part 5 in Figs. 7.2 and 7.14) contains six edit boxes, one check box, and one button. These allow the user to specify the different program options and are explained next.

1. *Design domain aspect ratio edit box:* This edit box  is used to specify the ratio of length (width) to height of the working design domain grid where the supports and loads of the structure are applied and also define the maximum width and height that the optimal structure can occupy. The default value is 1.5.
2. *Stress ratio edit box:* This edit box 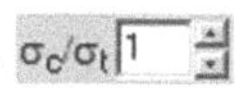 is used to specify the ratio between the compressive and tensile limit stresses. The default value is 1, which means that the same limit stress is used for both elements in tension and compression. If, for example a ratio of 0.3 is used, then the limit of the compressive stress is 0.3 that of the tensile stress. The range of values which can be used is $0.05 \leq \sigma_c/\sigma_t \leq 1$.
3. *Load magnitude edit box:* This edit box  is used to specify the magnitude of the applied load. The default value is 1, and for most classical optimization problems a value of 1 can be used. However the user can specify other values in the range of $0.1 \leq Load \leq 10$.
4. *Load angle edit box:* This edit box  is used to specify the angle of the applied load. The default value is $-90°$, which is measured with regards to the horizontal *X*-axis. The range of angles is $-180° \leq Angle \leq 180°$.
5. *Iterations edit box:* This edit box  is used to specify the maximum number of iterations before the program stops. The default value is 100 and the maximum number of iterations which can be specified is 200.
6. *Grid edit box:* This edit box 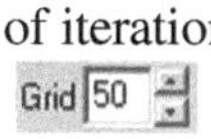 is used to specify the number of grid points along the horizontal *X*-axis. The default value is 50, but this can be changed to values ranging from 4 up to a maximum of 200 grid points along the horizontal *X*-axis.
7. *Symmetry check box:* This check box 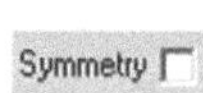 is used to force the program to impose symmetry on the solution if the loading and supports are symmetrical about either the horizontal or vertical axes. The default setting is not checked. If the structure is not symmetrical, then if this check box is selected, symmetry is not forced.
8. *Default button:* Selecting this button  resets all of the program options to their default values.

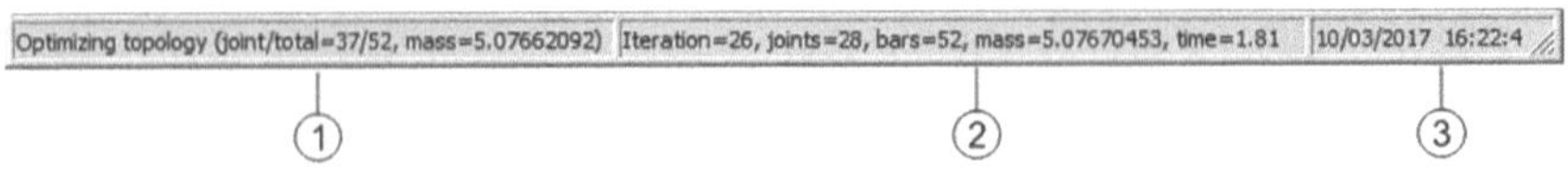

FIGURE 7.15 Program status bar.

7.2.2.6 Status Bar

The Status bar (part 6 in Figs.7.2 and 7.15) contains three cells where different information of the optimization process is displayed. These are explained next.

1. *Optimizing topology:* This cell shows information of how the mass of the structure improves at each optimization stage.
2. *Current optimal topology values:* This cell shows the current completed iteration number, the number of joints created in the structure and the number of bars connecting all of these joints. It also shows the current optimal mass and solution time in seconds.
3. *Current time and date:* This cell displays the current date and time.

7.3 STEP-BY-STEP GUIDE TO USE TTO

This section will guide the user through the necessary steps to generate a model and optimize it with TTO. It will also give information to the reader of the eleven *".dat"* files which correspond with the examples given in Chapter 5, Hands-On Applications of Structural Optimization.

7.3.1 Using TTO

There are seven steps needed to optimize a truss structure with TTO, these steps are:

1. Define the design aspect ratio of the design domain.
2. Specify the grid size.
3. Specify the boundary joints including their allowed displacement.
4. Specify the applied load, its magnitude, and direction.
5. Specify the maximum number of iterations.
6. Run the program.
7. View the results.

To show how to use TTO, the structure to be optimized is the Michell cantilever [2] of Section 5.2 with an aspect ratio of $\bar{L} = 1.82196$ (Fig. 7.16). As TTO basically optimized a non-dimensional structure and produces a non-dimensional optimal mass, then the dimensions are applied non-dimensionally and a unit load is all that is required. The seven steps mentioned above are explained next.

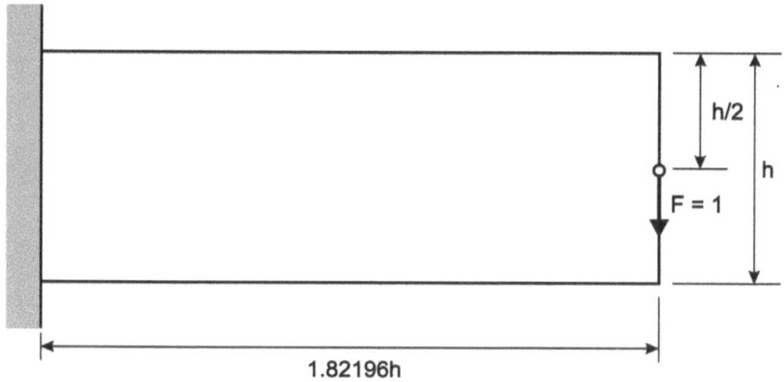

FIGURE 7.16 Michell cantilever of aspect ratio $\overline{L} = 1.82196$.

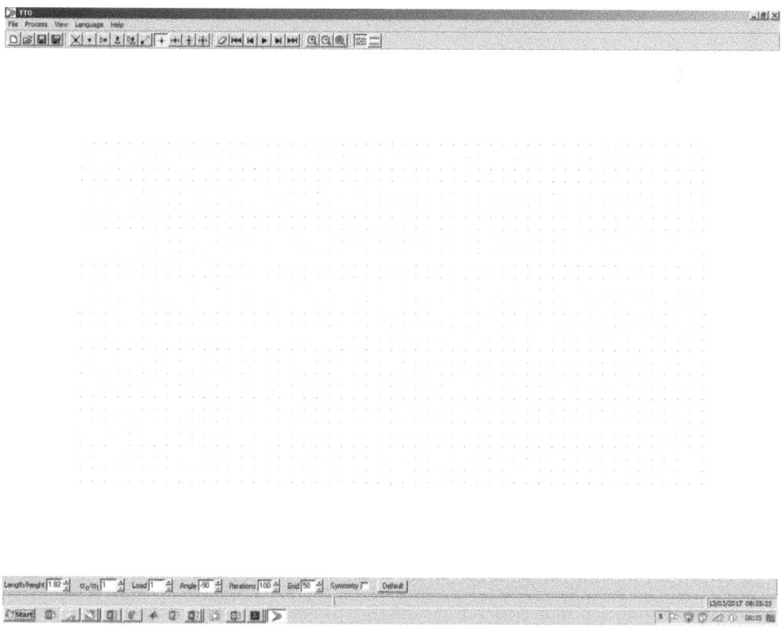

FIGURE 7.17 Michell cantilever working design domain grid of aspect ratio $\overline{L} = 1.82196$.

1. Start TTO.exe
2. In the *"Design domain aspect ratio edit box"* type the value *"1.82196"* and press Enter. This will automatically reset the design domain grid to this aspect ratio (Fig. 7.17).
3. Click over the "*Symmetry Check*" box, so that symmetry is checked. All other settings of the *"Program Options Bar"* have their default values.
4. Select the "*Add a Joint with all fixed displacements*" button and set it in the pressed down mode. Then click on both the top left grid point and then on the bottom left grid point to produce the supports of Fig. 7.18.

FIGURE 7.18 Application of the two supporting joints for the Michell cantilever.

5. Select the *"Add a Joint with an applied load"* button and set it in the pressed down mode. Then click on the central grid point on the right hand edge of the design domain grid to produce the applied load joint of Fig. 7.19.
6. Press the *"Optimize button"* and watch the structure being optimized.
7. When the optimization stops after 100 iterations, the final topology will be that of Fig. 7.20.

In Fig. 7.20, there is a set of two connecting bars which haven't quite reached the outer boundary of the structure in the henky-net region close to the fan line region to the left of the domain. If four more bars were added with the addition of four more joints on the top and bottom of the domain, this should happen. To achieve this now, change the maximum number of iterations to 104 in the *"Iterations Edit"* box and then press the *"Optimize button."* The emerging topology will now be complete (Fig. 7.21).

7.3.2 TTO Models in the Included Files

The eleven *".dat"* files which are included in the "TTOSoftware.zip" file correspond to the eleven examples of Chapter 5, Hands-On Applications of Structural Optimization. The examples associated with each TTO file name is given in Table 7.2.

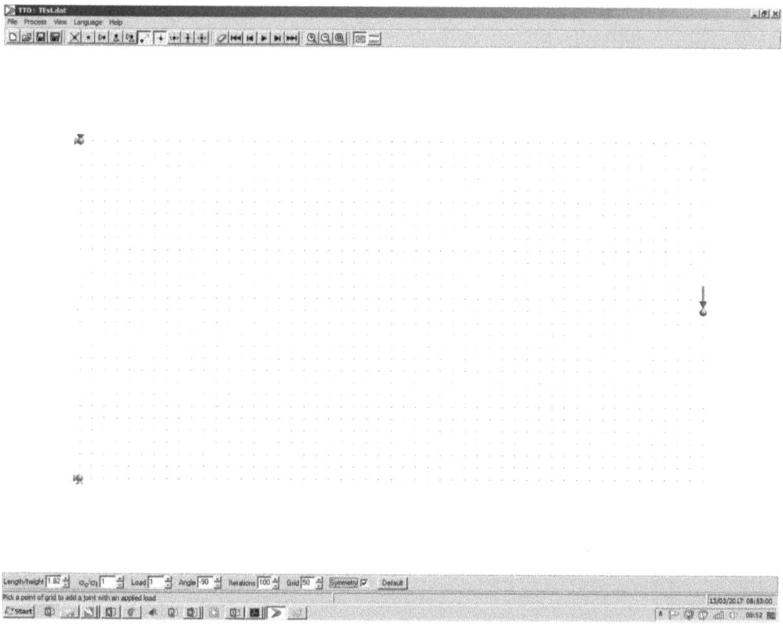

FIGURE 7.19 Application of the applied load joint for the Michell cantilever.

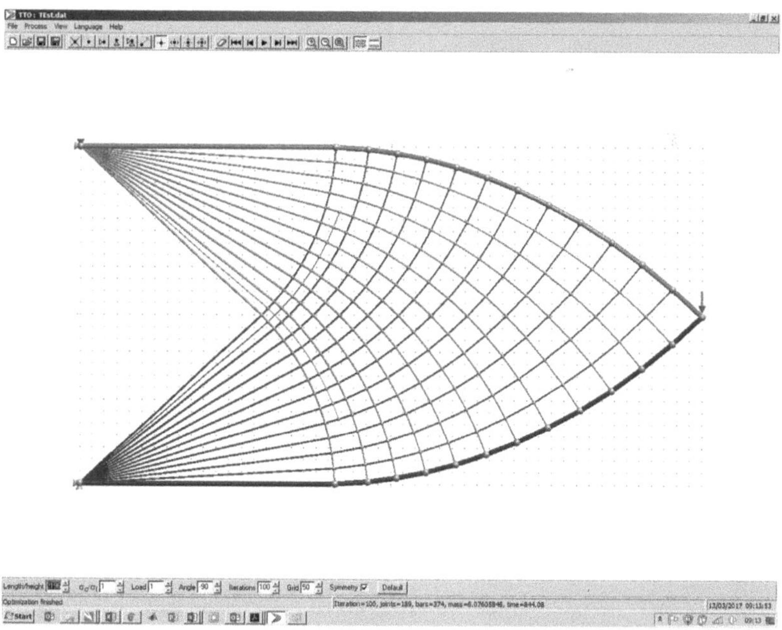

FIGURE 7.20 Optimal structural domain for the Michell cantilever after 100 iterations.

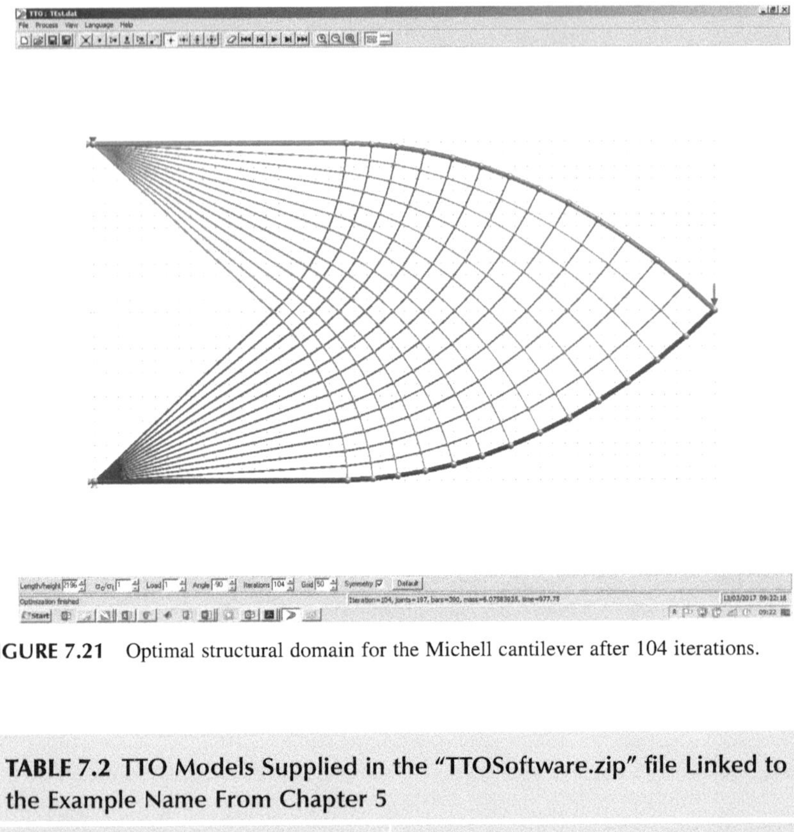

FIGURE 7.21 Optimal structural domain for the Michell cantilever after 104 iterations.

TABLE 7.2 TTO Models Supplied in the "TTOSoftware.zip" file Linked to the Example Name From Chapter 5

Example Name	File Name
Michell cantilever (Section 5.2) $\overline{L} = 0.5$	MichellCantilever_0.5.dat
Michell cantilever (Section 5.2) $\overline{L} = 1.82196$	MichellCantilever_1.82196.dat
Michell cantilever (Section 5.2) $\overline{L} = 3.35889$	MichellCantilever_3.35889.dat
MBB beam (Section 5.3) $\overline{L} = 2.40196$	MbbBeam_2.40196.dat
MBB beam (Section 5.3) $\overline{L} = 5.49846$	MbbBeam_5.49846.dat
Michell cantilever with fixed circular boundary (Section 5.4) $L/r = 10$	MichellCantilever_CircularBoundary_Lr10.dat
Michell cantilever with fixed circular boundary (Section 5.4) $L/r = 15$	MichellCantilever_CircularBoundary_Lr15.dat
Michell beam with fixed supports (Section 5.5)	MichellBeam_FixedSupports.dat

(Continued)

TABLE 7.2 (Continued)

Example Name	File Name
Michell beam with roller support (Section 5.6)	MichellBeam_RollerSupports.dat
Prager cantilever (Section 5.9): 12 iterations	PragerCantilever_12Iterations.dat
Prager cantilever (Section 5.9): 108 iterations	PragerCantilever_108Iterations.dat

7.4 SERA TOPOLOGY OPTIMIZATION PROGRAM

This section describes the MATLAB [1] implementation of the Sequential Element Rejection and Admission (SERA) method for topology optimization of structures and compliant mechanisms presented in Chapter 3, Discrete Method of Structural Optimization. The lines comprising this code include the definition of the design domain, finite element analysis, sensitivity analysis, mesh-independent filter, optimization algorithm and display of the results. Extensions and changes in the algorithm are also included in order to solve multiple load cases, active and passive elements and compliant mechanisms design.

To install the SERA program, please follow these steps:

1. Using Windows Explorer, create a directory called "\SERA" in your computer.
2. Please access the companion website for this book: https://www.elsevier.com/books-and-journals/book-companion/9780081009161 and please enter the "SERA software" link.
3. From the "SERA software" link download the zipped file "SeraSoftware.zip" into the SERA directory crated in your computer in step (1), and unzip the file in the same directory. The list of files is given in Table 7.3.
4. The main SERA program corresponds with the file named *"SERA.m"*, which is explained in Section 7.4.1. The other SERA programs are explained in Section 7.5.
5. Installation is now complete.
6. To run the SERA program(s), MATLAB needs to be active. Inside of the MATLAB Command window any of the MATLAB function calls of Eqs. (7.1, 7.2, 7.3, 7.4, 7.5, 7.7, 7.8, 7.9, 7.10, 7.11) need to be typed and executed.

The source code of the MATLAB program *"SERA.m"* is given and explained in Section 7.4.1. From the MATLAB Command window, the MATLAB function call of (7.1) needs to be executed in order to run the program.

TABLE 7.3 List of All Files in the "SERASoftware.zip" file from this Book Companion Website

Name	Size
SERA.m	5 kB
SERA_Compliant_Mechanism.m	6 kB
SERA_MBB.m	5 kB
SERA_Michell_Cantilever.m	5 kB
SERA_Multiple_Load_Cases.m	5 kB
SERA_Passive_Elements.m	6 kB

$$SERA(NelX,\ NelY,\ VolFrac,\ Rmin) \tag{7.1}$$

where: *SERA* is the name of the function and also the name of the "*SERA.m*" MATLAB file which consists of the program; *NelX* and *NelY* are the number of unit sized elements that form the horizontal and vertical sides respectively of the structure; *VolFrac* is the volume fraction and *Rmin* represents the size of the filter radius. There are additional variables which are defined inside of the source code and which can be edited if needed, these consist of: (1) Termination criteria; (2) Material model; (3) Boundary conditions; and (4) Plot style. The input line is similar to the style of the 99 line code developed by Sigmund [3] from which this program is based. The details of the MATLAB code and its modification to make it work for several examples are discussed in the following sections.

7.4.1 SERA Matlab Code

The SERA MATLAB program is given in the "*SERA.m*" file inside of the "SERASoftware.zip" file from this Book Companion Website. The source code is given in Table 7.4 where the default example structure to optimize is the Messerschmidt-Bölkow-Blohm (MBB) of Sections 5.3 and 7.5.1.

The main program starts by initializing the iteration number and the variable that takes account of the change in the objective function. Then a full rectangular design domain with *NelX* elements in the horizontal direction and *NelY* elements in the vertical direction is defined (initially all elements are "*real*" material). The Progression Ratio *PR*, the Smoothing Ratio *SR*, and the redistribution ratio *B* are set at the beginning of the main program as well. The main loop stars calculating the target volume fraction of the next iteration and the finite element subroutine is called in line 12. This function returns the displacement vector **U**, the global stiffness matrix *K* and the element stiffness matrix *Ke*, which is the same for all of the "*real*" material elements. The displacement vector and the global stiffness matrix are used to

TABLE 7.4 Listing of *"SERA.m"* Source Code

```
%%%%%%%%% SERA TOPOLOGY OPTIMIZATION CODE %%%%%%%%%
function SERA(NelX,NelY,VolFrac,Rmin)
% INITIALIZE
i = 1;Change = 1;
x(1:NelY,1:NelX) = 1.0;
Vol(i) = ceil(sum(sum(x)))/(NelX*NelY);
PR = 0.03;SR = 1.3;B = 0.003;
% START ITERATION
while Change > 0.001
   i = i + 1;
   Vol(i) = max(Vol(i-1)*(1-PR),VolFrac);
   [U,Ke,K] = FE(NelX,NelY,x);
   c(i) = 0.5*U'*K*U;
   for ely = 1:NelY;
     for elx = 1:NelX;
       n1 = (NelY + 1)*(elx-1) + ely;
       n2 = (NelY + 1)*elx + ely;
       Ue = U([2*n1-1;2*n1;2*n2-
       1;2*n2;2*n2 + 1;2*n2 + 2;2*n1 + 1;2*n1 + 2],1);
       alfa(ely,elx) = 0.5*Ue'*Ke*Ue;
     end
   end
 % FILTERING TECHNIQUE
   [alfa] = Filter(NelX,NelY,Rmin,x,alfa);
 % DESIGN UPDATE
   [x] = SERA_Update(NelX,NelY,alfa,x,Vol,VolFrac,i,SR,B);
   Vol(i) = ceil(sum(sum(x)))/(NelX*NelY);
 % CONVERGENCE CRITERION
   if i > 10
     Change = abs((sum(c(i-9:i-5))-sum(c(i-4:i)))/
     sum(c(i-4:i)));
   end
```

(Continued)

TABLE 7.4 (Continued)

```
% PRINT RESULTS
disp([' Iteration: ' sprintf('%4i', (i-1)) ' Volume fraction:'...
    sprintf('%6.3f', Vol(i)) ' Compliance: ' sprintf('%6.6f',c(i))])
% PLOT DENSITIES
  SeraMap = [0, 1, 1; 1, 1, 1]; colormap(SeraMap);
  imagesc(-x); axis equal; axis tight;
  axis off; pause(1e-6)
end
%%%%%%%%% FE-ANALYSIS %%%%%%%%%
function [U,Ke,K] = FE(NelX,NelY,x)
[Ke] = lk;
K = sparse(2*(NelX + 1)*(NelY + 1), 2*(NelX + 1)*(NelY + 1));
F = sparse(2*(NelY + 1)*(NelX + 1),1);
for elx = 1:NelX
  for ely = 1:NelY
  n1 = (NelY + 1)*(elx-1) + ely;
  n2 = (NelY + 1)*elx + ely;
  edof = [2*n1-1;2*n1;2*n2-1;2*n2;2*n2 + 1;2*n2 + 2;2*n1 + 1;2*n1 + 2];
  K(edof,edof) = K(edof,edof) + x(ely,elx)*Ke;
  end
end
% DEFINE LOADS AND SUPPORTS (FOR SPECIFIC PROBLEM - Default is MBB Beam)
F(2,1) = -1;
fixeddofs = union([1:2:2*(NelY + 1)],[2*(NelX + 1)*(NelY + 1)]);
alldofs = [1:2*(NelY + 1)*(NelX + 1)];
freedofs = setdiff(alldofs,fixeddofs);
% SOLVING
U = zeros(2*(NelY + 1)*(NelX + 1),1);
```

(Continued)

TABLE 7.4 (Continued)

```
U(freedofs,:)=K(freedofs,freedofs) \ F(freedofs,:);
U(fixeddofs,:)=0;
%%%%%%%%% ELEMENT STIFFNESS MATRIX %%%%%%%%%
function [Ke]=lk
E=1; nu=0.3;
k=[1/2-nu/6;1/8+nu/8;-1/4-nu/12;-1/8+3*nu/8;
-1/4+nu/12;-1/8-nu/8;nu/6;1/8-3*nu/8];
Ke=E/(1-nu^2)*...
    [k(1),k(2),k(3),k(4),k(5),k(6),k(7),k(8);
k(2),k(1),k(8),k(7),k(6),k(5),k(4),k(3);
k(3),k(8),k(1),k(6),k(7),k(4),k(5),k(2);
k(4),k(7),k(6),k(1),k(8),k(3),k(2),k(5);
k(5),k(6),k(7),k(8),k(1),k(2),k(3),k(4);
k(6),k(5),k(4),k(3),k(2),k(1),k(8),k(7);
k(7),k(4),k(5),k(2),k(3),k(8),k(1),k(6);
k(8),k(3),k(2),k(5),k(4),k(7),k(6),k(1)];
%%%%%%%%% MESH-INDEPENDENCY FILTER %%%%%%%%%
function [alfaNew]=Filter(NelX,NelY,Rmin,x,alfa)
alfaNew=zeros(NelY,NelX);
for i=1:NelX
  for j=1:NelY
    sum=0.0;
    for k=max(i-floor(Rmin),1):min(i+floor(Rmin),NelX)
      for l=max(j-floor(Rmin),1):min(j+floor(Rmin),NelY)
        sum=sum+max(0,Rmin-sqrt((i-k)^2+(j-l)^2));
        alfaNew(j,i)=alfaNew(j,i)+...
          max(0,Rmin-sqrt((i-k)^2+(j-l)^2))*x(l,k)*
          alfa(l,k);
      end
    end
    alfaNew(j,i)=alfaNew(j,i)/sum;
```

(Continued)

TABLE 7.4 (Continued)

```
    end
  end
  %%%%%%%%% SERA UPDATE %%%%%%%%%
  function [x] = SERA_Update(NelX,NelY,alfa,x,Vol,VolFrac,i,SR,B);
  alfa_min = min(min(alfa));alfa_max = max(max(alfa));
  alfa_V(1:NelY,1:NelX) = alfa_min;alfa_R(1:NelY,1:NelX) = alfa_max;
  for ely = 1:NelY
    for elx = 1:NelX
      if x(ely,elx) > 1e-3
        alfa_R(ely,elx) = alfa(ely,elx);
      else
          alfa_V(ely,elx) = alfa(ely,elx);
      end
    end
  end
  if Vol(i) > VolFrac
    DeltaV(i) = abs(Vol(i)-Vol(i-1));
    DeltaV_Rem = DeltaV(i)*(SR);
    NumElem_Rem = max(1,floor(NelX*NelY*DeltaV_Rem));
    [x,NumElem_Rem] = Update_R(NelX,NelY,x,alfa_R,NumElem_Rem);
    if i > 2
      NumElem_Add = max(1,floor(NumElem_Rem*(SR-1)));
      [x] = Update_V(NelX,NelY,x,alfa_V,NumElem_Add);
    end
  else DeltaV_Rem = B*VolFrac;
    NumElem_Rem = max(1,floor(NelX*NelY*DeltaV_Rem));
    [x,NumElem_Rem] = Update_R(NelX,NelY,x,alfa_R,NumElem_Rem);
    NumElem_Add = NumElem_Rem;
    [x] = Update_V(NelX,NelY,x,alfa_V,NumElem_Add);
```

(Continued)

TABLE 7.4 (Continued)

```
end
%%%%%%%%% UPDATE_R %%%%%%%%%
function[x,NumElem_Rem] = Update_R(NelX,NelY,x,alfa_R,NumElem_Rem)
alfa_R_vec = sort(reshape(alfa_R,(NelX*NelY),1),'descend');
alfa_R_th = alfa_R_vec((NelX*NelY)-NumElem_Rem,1);
NumElem_Rem = 0;
for ely = 1:NelY
  for elx = 1:NelX
    if x(ely,elx) == 1
      if ((alfa_R(ely,elx)-alfa_R_th)/alfa_R_th) < 1e-4
        x(ely,elx) = 1e-3;
        NumElem_Rem = NumElem_Rem + 1;
      end
    end
  end
end
%%%%%%%%% UPDATE_V %%%%%%%%%
function [x] = Update_V(NelX,NelY,x,alfa_V,NumElem_Add)
alfa_V_vec = sort(reshape(alfa_V,(NelX*NelY),1),'descend');
alfa_V_th = alfa_V_vec(NumElem_Add,1);
for ely = 1:NelY
  for elx = 1:NelX
    if x(ely,elx) == 1e-3
      if ((alfa_V(ely,elx)-alfa_V_th)/alfa_V_th) > -1e-4
        x(ely,elx) = 1.;
    end
    end
  end
end
```

determine the objective function *c(i)* in line 13. The element displacement vector *Ue* is extracted from the global displacement vector *U* in line 18, and this is used to calculate the sensitivity numbers for all elements *alfa(ely,elx)* in line 19. Following this loop, there is a call to the mesh-independency filter in line 23 and the SERA optimization algorithm in line 25. The new volume fraction for the current iteration is calculated in line 26, and the normalized change in the value of the objective function is calculated in line 29. The intermediate results are printed and plotted in lines 32–37. The steps mentioned above are repeated unless the optimization loop is terminated. This is achieved when the relative change of the objective function in the last 10 iterations is less than 0.001, which corresponds which a change of 0.1%, and takes place in the *while* statement of line 9. The variable *Change* used for this comparison is calculated line 29.

The optimization loop begins in line 25 with a call to the *SERA_Update* function. The algorithm for this function starts in line 92, and immediately sorts the "*real*" and "*virtual*" elements, which are stored in two different matrixes *alfa_R* and *alfa_V* respectively. The next step depends on the actual volume fraction of the design. If the *"real"* material volume fraction is larger than the specified volume fraction *VolFRAC*, the new target volume fraction *DeltaV(i)* is calculated as well as the number of elements to be removed *NumElem_Rem* in line 107. If the iteration counter *i* is greater than 2 the number of elements to add *NumElem_Add* is then calculated in line 110. When the "*real*" material target volume fraction is reached in line 113, the material redistribution process takes place. In this step the same amount of material is added and rejected, calculated in lines 113 to 118. The subroutine which controls the amount of material to be removed is called *Update_R* and starts in line 120 and the subroutine which controls the amount of material to be introduced is called *Update_V* and starts in line 135. *Update_R* sorts elements with *"real"* material according to their sensitivity number from highest to lowest, and defines the threshold value *alfa_R_th* in line 122, which depends on the number of elements that should be removed. "*Real*" elements are transformed into "*virtual*" elements if their sensitivity is lower than the previously defined threshold value. In order to avoid numerical issues and unsymmetrical results, this condition is specified in relative terms in line 127. *Update_V* works in a similar manner, by transforming "*virtual*" into *"real"* elements' when their sensitivity number is higher than the corresponding threshold value *alfa_V_th*, calculated in line 137. The optimization loop terminates when the convergence criteria is met, which means that the material redistribution stage cannot improve any further the value of the objective function.

The code requires additional lines to incorporate problems with different boundary conditions, multiple load cases, passive elements and compliant mechanisms. Section 7.5 shows the reader how these changes can be implemented to the *"SERA.m"* program. This section also shows the effect that changes to some of the SERA parameters have on the emerging topologies.

7.5 MODIFYING THE SERA CODE TO SOLVE DIFFERENT EXAMPLES

Five different examples are presented here to show how the *"SERA.m"* MATLAB code can be easily manipulated to solve them. These examples are the: (1) MBB Beam; (2) Michell cantilevers optimized for minimum compliance; (3) Multiple load case short cantilever; (4) Short cantilever with a nondesign domain region; and (5) Inverted compliant mechanism.

7.5.1 The Messerschmidt-Bölkow-Blohm Beam (MBB)

The SERA MATLAB program was used to optimize the Messerschmidt-Bölkow-Blohm (MBB) beam [2], a full description of this example is given in Section 5.3. Although the *"SERA.m"* program default structure is the MBB beam, this program was copied to the filename *"SERA_MBB.m"*. As the SERA program uses unit sized square finite elements of unit depth, the closest dimensioned structure to the MBB beam of aspect ratio $\overline{L} = 2.40196$ is a beam of dimensions 96 x 40 elements for the half structural domain of Fig. 7.22, corresponding to an aspect ratio of $\overline{L} = 96/40 = 2.4$.

The volume fraction limit used was 62.5% and two different filter radii were used: 1.5 and 3.0. The MATLAB function calls for these two cases are Eqs. (7.2) and (7.3), respectively. The topologies which emerge from these two inputs are given in Fig. 7.23A and B.

$$SERA\_MBB(96, 40, 0.625, 1.5) \tag{7.2}$$

$$SERA\_MBB(96, 40, 0.625, 3.0) \tag{7.3}$$

7.5.2 The Michell Cantilever

The SERA MATLAB program was used to optimize the Michell cantilever [2] of Section 5.2. The *"SERA.m"* program of Section 7.4.1 needs to be modified

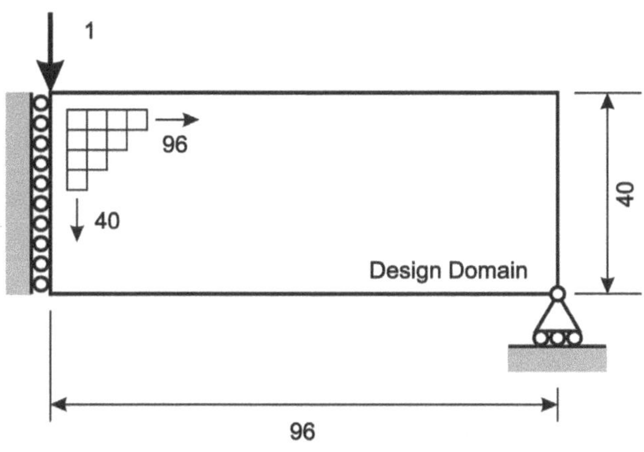

FIGURE 7.22 MBB beam design domain used in the SERA MATLAB program.

FIGURE 7.23 MBB beam design for a volume fraction of 0.625 and two filter radii: (A) 1.5 and (B) 3.0.

to account for the different boundary condition which corresponds with the left hand edge being fully fixed. The loading is also different with the vertical force applied in the middle of the right hand edge (Fig. 5.1). These changes consist of replacing the current lines 53 and 54 of the program:

```
53 F(2,1) = -1;
54 fixeddofs = union([1:2:2*(NelY + 1)],[2*(NelX + 1)*(NelY + 1)]);
```

With the following two new lines:

```
53 F(2*(NelY + 1)*(NelX + 1)-NelY,1) = -1.0;
54 fixeddofs = [1:2*(NelY + 1)];
```

The resulting program is included inside of the "SERASoftware.zip" file from this Book Companion Website and is called *"SERA_Michell_Cantilever.m"*.

The volume fraction limit used was 65% and the filter radius used was 1.3. The MATLAB function call for the two Michell cantilevers of aspect ratio: (a) $\overline{L} = 1.82196$ and (b) $\overline{L} = 3.35889$ [2], are given in Eqs. (7.4) and (7.5), respectively. Since in the program only uses square elements of unit size, the closest aspect ratios for the two cases are (a) $\overline{L} = 168/92 = 1.8261$ and (b) $\overline{L} = 168/50 = 3.36$. The topologies which emerge from these two inputs are given in Fig. 7.24A and B.

$$SERA\_Michell\_Cantilever(168, 92, 0.65, 1.3) \tag{7.4}$$

$$SERA\_Michell\_Cantilever(168, 50, 0.65, 1.3) \tag{7.5}$$

7.5.3 Multiple Load Case Problem

The *"SERA.m"* program of Section 7.4.1 needs to be modified to account for problems with multiple load cases. To demonstrate how the modification of

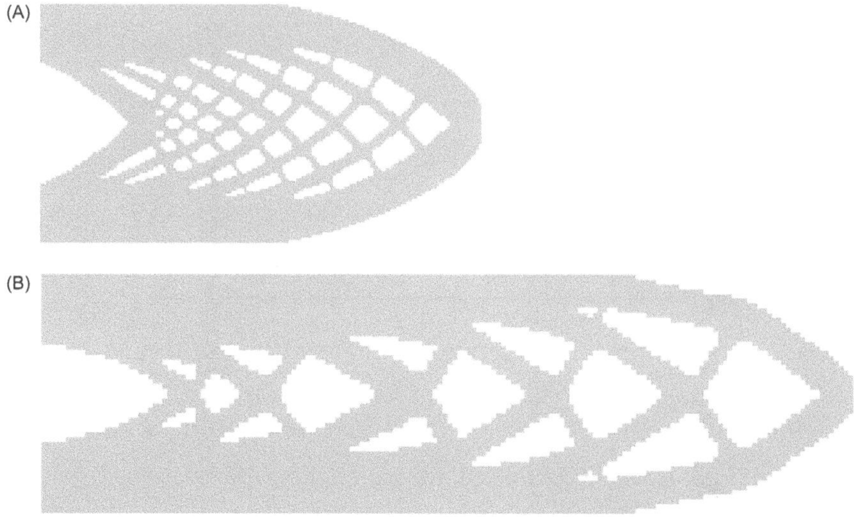

FIGURE 7.24 Michell cantilever design for a set volume fraction of 0.65 and two aspect ratios: (A) $\overline{L} = 1.8261$ and (B) $\overline{L} = 3.36$.

this program is done, two-load cases are considered. The force and displacement vectors defined in the current lines 43 and 58:

```
43  F = sparse(2*(NelY + 1)*(NelX + 1),1);
58  U = zeros(2*(NelY + 1)*(NelX + 1),1);
```

Must be replaced with the following new two-column vectors (lines 43 and 58):

```
43  F = sparse(2*(NelY + 1)*(NelX + 1),2);
58  U = zeros(2*(NelY +1)*(NelX +1),2);
```

As explained in Section 3.3.3, the objective function is defined as the sum of the compliances for each of the load cases, given by Eq. (7.6).

$$c(\rho) = \sum_{k=1}^{K} \omega_k \left[\mathbf{U}_k^T \mathbf{K} \mathbf{U}_k\right], \quad where: \sum_{k=1}^{K} \omega_k = 1 \tag{7.6}$$

where: K is the number of load cases; ω_k is the weighting factor for each load case, which in this instance is equal for each load case, such that: $\omega_1 = \omega_2 = 0.5$.

This requires the modification of the current line 13:

```
13  c(i) = 0.5*U'*K*U;
```

Which needs to be replaced with this new line 13:

```
13  c(i) = 0.5*U(:,1)'*K* U(:,1)+ 0.5*U(:,2)'*K* U(:,2);
```

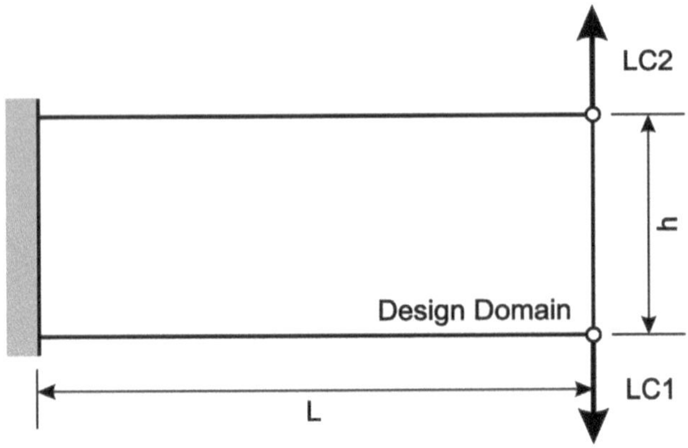

FIGURE 7.25 Rectangular beam design domain used in the multiple load case problem.

As two load cases are considered, the current lines 18–19:

```
18 Ue = U
   ([2*n1–1;2*n1;2*n2–1;2*n2;2*n2 + 1;2*n2 + 2;2*n1 + 1;2*n1 + 2],1);
19 alfa(ely,elx) = 0.5*Ue'*Ke*Ue;
```

Must be replaced with the following five new line numbers 18 to 22.

```
18 alfa(ely,elx) = 0.0;
19 for j = 1:2
20 Ue = U([2*n1–1;2*n1;
   2*n2–1;2*n2;2*n2 + 1;2*n2 + 2;2*n1 + 1;2*n1 + 2],j);
21 alfa(ely,elx) = alfa(ely,elx) + Ue'*Ke*Ue;
22 end
```

Note that as 3 extra lines have just been introduced into the program, all line numbers in the original program of Section 7.4.1 *from line 20 onwards, will have their line number increased by 3. All subsequent changes to the program refer to these new line numbers.*

To solve the two-load case problem of the rectangular cantilever beam of Fig. 7.25, the loading condition of the new line 56 and supports of the new line 57:

```
56 F(2,1) = -1;
57 fixeddofs = union([1:2:2*(NelY + 1)],[2*(NelX + 1)*(NelY + 1)]);
```

Must be replaced with these three new line numbers 56 to 58:

```
56 F(2*(NelX +1)*(NelY +1),1) = -1.0;
57 F(2*(NelX)*(NelY +1) + 2,2) = 1.0;
58 fixeddofs = [1:2*(NelY + 1)];
```

The resulting program is included inside of the "SERASoftware.zip" file from this Book Companion Website and is called *"SERA_Multiple_Load_Cases.m"*.

Two problems are presented: (a) A square cantilever with a volume fraction of 40% and a filter radius of 1.5 [3] and (b) a rectangular cantilever of aspect ratio 2 with a volume fraction of 50% and a filter radius of 1.3. The MATLAB function call for both problems is given in Eqs. (7.7) and (7.8),

(A)

(B)

FIGURE 7.26 Rectangular cantilever designs for two load cases and aspect ratios: (A) $\overline{L}=1$ and (B) $\overline{L}=2$.

respectively. The topologies which emerge from these two inputs are given in Fig. 7.26A and B.

$$SERA\_Multiple\_Load\_Cases\,(80, 80, 0.4, 1.5) \quad (7.7)$$

$$SERA\_Multiple\_Load\_Cases\,(160, 80, 0.5, 1.3) \quad (7.8)$$

7.5.4 Structures with Passive Elements

There may be instances when part of a structure should not take part in the optimization, or there is a region which needs to be kept void permanently. This can be achieved by extending the "*SERA.m*" program of Section 7.4.1 to incorporate *Passive* elements. This is achieved by defining a passive element array of size NelY by NelX, *passive(NelY, NelX)*. All passive elements are allocated the value 1, and all elements which can take part in the optimization are given the value 0. The way to introduce passive elements to the "*SERA.m*" program is as follows.

Replace the existing line 5 which consists of:

```
5  x(1:NelY,1:NelX)=1.0;
```

With the following new line 5, which adds the passive array of dimension NelY, NelX and sets all values to 0.

```
5 x(1:NelY,1:NelX)=1.0; passive(NelY, NelX)=0;
```

The Passive elements need to be passed to the subroutine SERA_Update, so the current line 92:

```
92 function [x]=SERA_Update(NelX,NelY,alfa,x,Vol,VolFrac,i,SR,B);
```

Needs to be replaced with:

```
92 function [x]=SERA_Update(NelX,NelY,alfa,x,Vol,VolFrac,i,SR,
   B,passive);
```

In order for void passive elements to never have their "*virtual*" status changed to "*real*," the value of their sensitivity numbers are given the minimum sensitivity value *alfa_min*. These elements are then moved to the bottom of the *alfa_V array* and are therefore not selected to be changed from "*virtual*" to "*real*" material. Alternatively, for non-design domain material which needs to have the same property as the rest of the "*real*" structure, the value of their sensitivity numbers are given the maximum sensitivity value *alfa_max*. These elements are then moved to the top of the *alfa_R array* and are therefore not selected to be changed from "*real*" to "*virtual*" material. This requires that the code of lines 95–103 below:

```
95  for ely=1:NelY
96      for elx=1:NelX
97          if x(ely,elx)>1e-3
98              alfa_R(ely,elx)=alfa(ely,elx);
99          else
100             alfa_V(ely,elx)=alfa(ely,elx);
101         end
102     end
103 end
```

Is replaced with the following 15 lines of code for the new line numbers 95 to 109:

```
95  for ely=1:NelY
96      for elx=1:NelX
97          if x(ely,elx)>1e-3
98                alfa_R(ely,elx)=alfa(ely,elx);
99                if (passive(ely,elx)>0)
100                  alfa_R(ely,elx)=alfa_max;
101               end
102         else
103            alfa_V(ely,elx)=alfa(ely,elx);
104            if (passive(ely,elx)>0)
105                  alfa_V(ely,elx)=alfa_min;
106               end
107         end
108     end
109 end
```

Having updated the subroutine and passed the passive elements to it, where the subroutine is called in line 25 of the code, also needs to be updated. The current line 25, which has the following code:

```
25 [x]=SERA_Update(NelX,NelY,alfa,x,Vol,VolFrac,i,SR,B);
```

Needs to be replaced by the following new line 25, where information about passive elements are passed to it.

```
25 [x]=SERA_Update(NelX,NelY,alfa,x,Vol,VolFrac,i,SR,B,passive);
```

In order to add a passive circular design domain to a structure where its center is in location (NelY/3, NelX/3) and where its radius is equal to NelY/4, the following 9 lines of code have to be inserted between the current lines 5 and 6, so that the new lines 6 to 14 become:

```
6  Centre_x=NelX/3; Centre_y=NelY/3; Radius=NelY/4;
7  for ely=1:NelY
8      for elx=1:NelX
9          if sqrt((ely-Centre_y)^2+(elx-Centre_x)^2)<Radius;
10             passive(ely,elx)=1;
11             x(ely,elx)=1e-4;
12         end
13     end
14 end
```

The new line 11 above, is where the type of passive element is specified. If the value of the density for the passive element is equal to 10e-4, it is classified as void. If on the other hand, it has a value equal to 1, it is classified as a non-design domain which therefore has the same properties as the rest of the structure but is not optimized.

To show the effect of passive elements on a structural domain, the Michell cantilever example of Ref. [2] of Section 7.5.2 with $\overline{L}=$ 1.8261 is used, this will need lines 62 and 63 below:

```
62 F(2,1)=-1;
63 fixeddofs=union([1:2:2*(NelY+1)],[2*(NelX+1)*(NelY+1)]);
```

Replaces with the following two new lines 62 and 63:

```
62 F(2*(NelY+1)*(NelX+1)-NelY,1)=-1.0;
63 fixeddofs=[1:2*(NelY+1)];
```

The resulting program is included inside of the "SERASoftware.zip" file from this Book Companion Website and is called *"SERA_Passive_Elements.m"*.

The volume fraction limit used was 50% with a filter radius of 1.3. Two different types of passive elements were tested, void, and nondesign domain. For the void example, line 11 needs to be:

```
11 x(ely,elx)=1e-4;
```

The MATLAB call for this case is Eq. (7.9). The emerging topology is given in Fig. 7.27A.

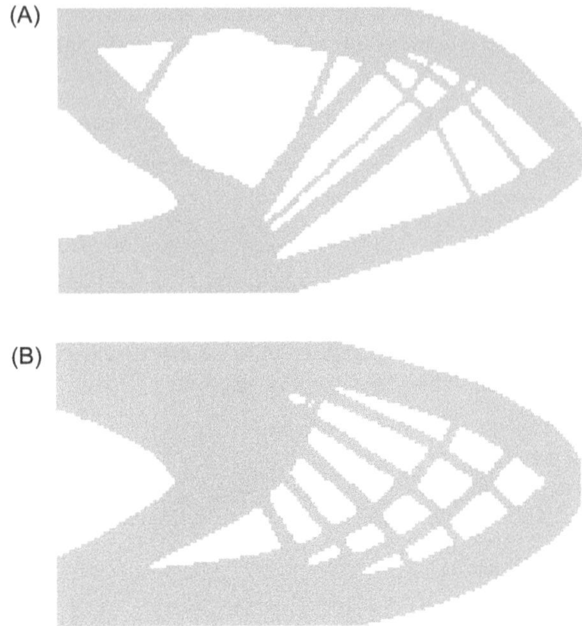

FIGURE 7.27 Michell cantilevers of aspect ratio $\overline{L} = 1.826$ with circular (A) Void passive element and (B) Non-design domain passive element.

$$SERA\_Passive\_Elements(168, 92, 0.5, 1.3) \tag{7.9}$$

For the case of using non-design domain passive elements, line 11 needs to be:

```
11  x(ely,elx)=1;
```

The MATLAB call for this case is Eq. (7.10). The emerging topology is given in Fig. 7.27B.

$$SERA\_Passive\_Elements(168, 92, 0.65, 1.3) \tag{7.10}$$

7.5.5 Compliant Mechanism Problems

The way to modify the "*SERA.m*" program of Section 7.4.1 to optimize compliant mechanisms is as follows. Compliant problems require the progression, smoothing, and material redistribution ratios to be adjusted to stabilize and improve convergence of the problem. The existing values set in the line 7 below:

```
7  PR=0.03;SR=1.3;B=0.003;
```

Needs to be changed to these new values of this new line 7:

```
7  PR=0.04;SR=1.4;B=0.003;
```

A compliant mechanism optimum design involves two load cases: (1) An input load case and (2) a dummy (or adjoint) load case. The allocation of the force and displacement vectors for the "*real*" and "*adjoint*" load cases is similar to the two load case problem of Section 7.5.3. But instead of calculating the compliance of the structure, the Mutual Potential Energy (MPE) needs to be calculated (see Section 3.3.4). This requires line 13, which has the following code:

```
13  c(i)=0.5*U'*K*U;
```

To be replaced with this new line 13 code:

```
13  Uout(i)=U(:,1)'*K*U(:,2);
```

The sensitivity values are obtained in terms of the solutions to the "*real*" and "*adjoint*" load case. These correspond to the first and second column of the displacement matrix U. This requires the current lines 18 and 19:

```
18  Ue=U
    ([2*n1-1;2*n1;2*n2-1;2*n2;2*n2+1;2*n2+2;2*n1+1;2*n1+2],1);
19  alfa(ely,elx)=0.5*Ue'*Ke*Ue;
```

To be replaced with the following new three lines, 18–20:

```
18  Ue1=U([2*n1-1;2*n1;
    2*n2-1;2*n2;2*n2+1;2*n2+2;2*n1+1;2*n1+2],1);
19  Ue2=U([2*n1-1;2*n1;
    2*n2-1;2*n2;2*n2+1;2*n2+2;2*n1+1;2*n1+2],2);
20  alfa(ely,elx)=-Ue1'*Ke*Ue2;
```

Note that as 1 extra line was introduced into the program, all line numbers in the original program of Section 7.4.1 from line 19 onwards, will have their line number increased by 1. All subsequent changes to the program refer to these new line numbers.

The line where the convergence criterion is calculated needs to be changed from compliance to MPE. So the command of the new line number 30:

```
30  Change=abs((sum(c(i-9:i-5))-sum(c(i-4:i)))/sum(c(i-4:i)));
```

Needs to be changed to:

```
30  Change=abs((sum(Uout(i-9:i-5))-sum(Uout(i-4:i)))/sum(Uout
    (i-4:i)));
```

The line where the results are displayed, also need to be changed from compliance to MPE. So the command of the new line number 34:

```
34  sprintf('%6.3f', Vol(i)) ' Compliance: ' sprintf('%6.6f',
    c(i))])
```

Needs to be changed to:

```
34  sprintf('%6.3f', Vol(i)) ' MPE: ' sprintf('%6.6f', Uout(i))])
```

The force and displacement vectors defined for a single load case in the new lines 44 and 59:

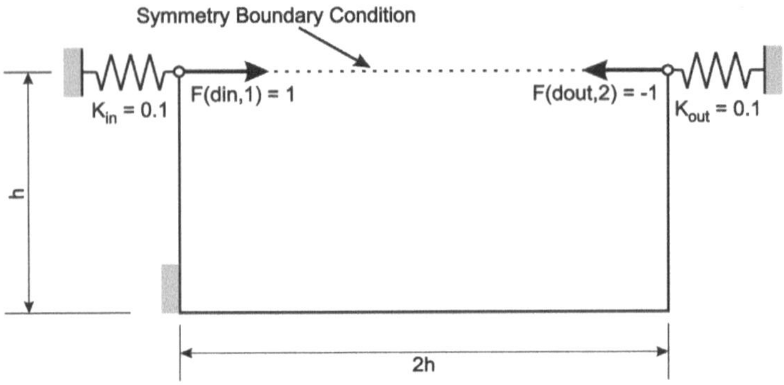

FIGURE 7.28 Half design domain used for the inverter compliant mechanism problem.

```
44 F = sparse(2*(NelY + 1)*(NelX + 1),1);
59 U = zeros(2*(NelY + 1)*(NelX + 1),1);
```

Must be replaced with the following new two-column vectors of new lines 44 and 59:

```
44 F = sparse(2*(NelY + 1)*(NelX + 1),2);
59 U = zeros(2*(NelY + 1)*(NelX + 1),2);
```

The sensitivity number *alfa_R_th* and *alfa_V_th* used in the denominator of the functions *Update_R* and *Update_V* can have either positive of negative values, depending on the desired direction of the compliant mechanics. For the elements to be appropriately selected in order to become "*real*" or "*virtual*," these denominator terms need to always be positive. This means that new lines 128 and 142 below:

```
128 if ((alfa_R(ely,elx)-alfa_R_th)/alfa_R_th) < 1e-4
142 if ((alfa_V(ely,elx)-alfa_V_th)/alfa_V_th) > -1e-4
```

Need to be changed to these new lines 128 and 142:

```
128 if ((alfa_R(ely,elx)-alfa_R_th)/abs(alfa_R_th)) < 1e-4
142 if ((alfa_V(ely,elx)-alfa_V_th)/abs(alfa_V_th)) > -1e-4
```

The specific compliant mechanism problem loading and boundary conditions can now be inserted to the program. This example consists of a half model of the inverter mechanism of Section 5.10 (Fig. 7.28). The external springs are added with a default value of 0.1 to the input and output points of the design domain. The input and output port degrees of freedom are called d_{in} and d_{out}, respectively.

The boundary conditions consists of a fixed location at the bottom left hand corner and of a symmetry boundary condition (*Y*-displacement fixed, *X*-displacement allowed) along the top section of the structure. To apply these conditions, new lines 54 and 55 of the code:

```
54 F(2,1) = -1;
55 fixeddofs = union([1:2:2*(NelY + 1)],[2*(NelX + 1)*(NelY + 1)]);
```

FIGURE 7.29 Inverter compliant mechanism design.

Need to be replaced with the following new 8 lines, 54–61.

```
54 din=1;
55 dout=2*NelX*(NelY+1)+1;
56 K(din,din)=K(din,din)+0.1;
57 K(dout,dout)=K(dout,dout)+0.1;
58 F(din,1)=1;
59 F(dout,2)=-1;
60 fixeddofs=union([2:2*(NelY+1):2*(NelY+1)*(NelX+1)],...
61     [2*(NelY+1):-1:2*(NelY+1)-3]);
```

The resulting program is included inside of the "SERASoftware.zip" file from this Book Companion Website and is called *"SERA_Compliant_Mechanism.m"*.

The volume fraction limit used was 30% with a radius of 1.1. The MATLAB function call for this case is Eq. (7.11). The emerging topology is given in Fig. 7.29.

$$SERA\_Compliant\_Mechanism(120, 60, 0.3, 1.1) \tag{7.11}$$

7.6 ISOLINES TOPOLOGY DESIGN PROGRAM (liteITD)

The liteITD program is a fully interactive topology design GUI developed using MATLAB [1]. It consists of three internal modules: (1) Preprocessor (geometric modeling, material properties, loading conditions, and supports); (2) Solution (Finite Element (FE) mesh generator and FE Analysis (FEA) solver); and (3) Optimizer (topology optimization algorithms and visualization options).

The liteITD program (version 1.0) implements the von Mises stress option for the topology optimization of two-dimensional continuum structures with: (1) Single and (2) Multiple loading conditions; (3) Different material properties/behaviour in tension and compression; and (4) Multimaterials. The liteITD program is included inside of the "liteITDSoftware.zip" file from this Book Companion Website.

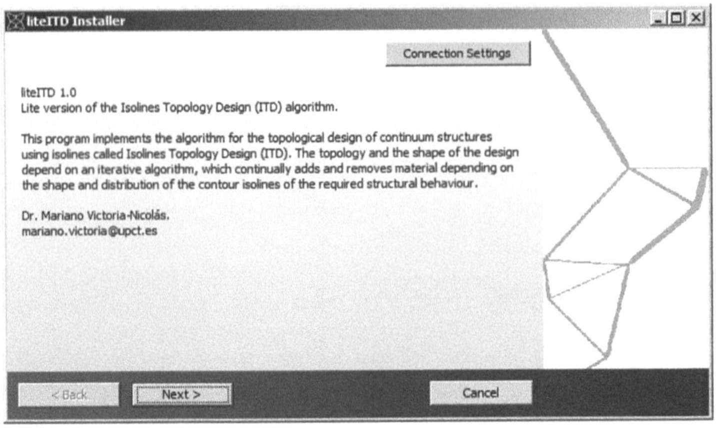

FIGURE 7.30 The installation window for liteITD.

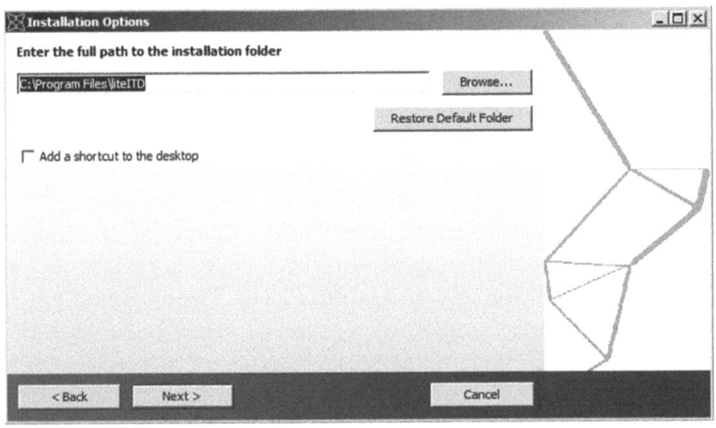

FIGURE 7.31 Installation options menu.

7.6.1 System Requirements and Installation of liteITD Software

The liteITD program was generated using MATLAB 2014a [1], and it is a 64-bit application for the Windows operating system. To install liteITD, please carry out the following steps:

1. Using Windows Explorer, create a directory called "\liteITD" in your computer. Please access the companion website for this book: https://www.elsevier.com/books-and-journals/book-companion/9780081009161, enter the "liteITD software" link, download the zipped file "liteITDSoftware.zip" into the liteITD directory created in your PC and unzip this file in the same directory.
2. Double-click on the liteITD installer "MyAppInstaller_mcr.exe". This will launch the Installer menu of Fig. 7.30.

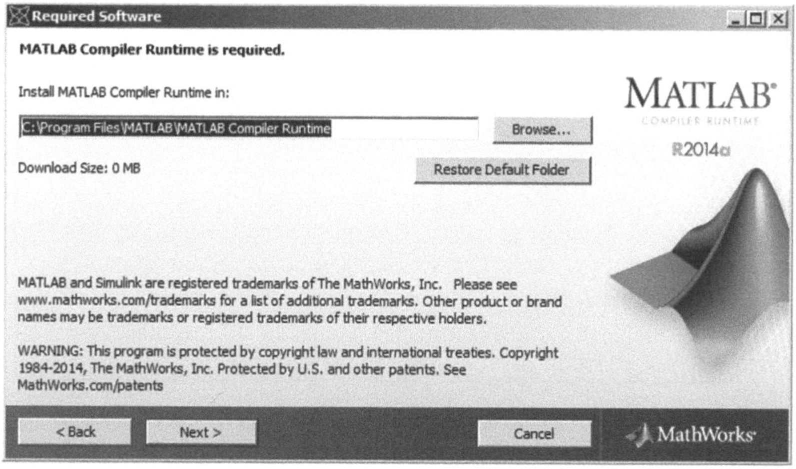

FIGURE 7.32 Menu to install the required MATLAB compiler.

3. Click Next to advance to the installation options menu (Fig. 7.31). Check that the appropriate directory path for the installation of liteITD has been set correctly.
4. Click Next to advance to the menu to install the required MATLAB compiler (Fig. 7.32). If asked about creating the destination folder, select Yes. If the user already has the appropriate version of the MATLAB Compiler Runtime (MCR) standalone set of shared libraries installed on the system, this page will have a message indicating that the user does not have to install the MCR. If the user receives this message, go straight to step 8.
5. Click Next to advance to the license agreement page. If asked about creating the destination folder, select Yes.
6. Read the license agreement and check Yes to accept the license.
7. Click Next to advance to the confirmation page.
8. Click Install. The installer then installs liteITD.
9. Click Finish.
10. Check that a liteITD icon has been created on the desktop. If one was created, go to step 12, if not, then go to step 11.
11. Go to *Start- >All Programs*, and find liteITD and click on it with the right mouse button. Then select *Send to - > Desktop (create shortcut).*
12. You can now run liteITD from the Desktop.

Note that when you run liteITD for the first time, it should create a directory on the Desktop called outITD, which is the default directory for all running files to be stored.

7.6.2 Overview of the liteITD Interface

When liteITD is started, the graphical interface of Fig. 7.33 appears. This is similar to most Windows applications as it has most of the usual features.

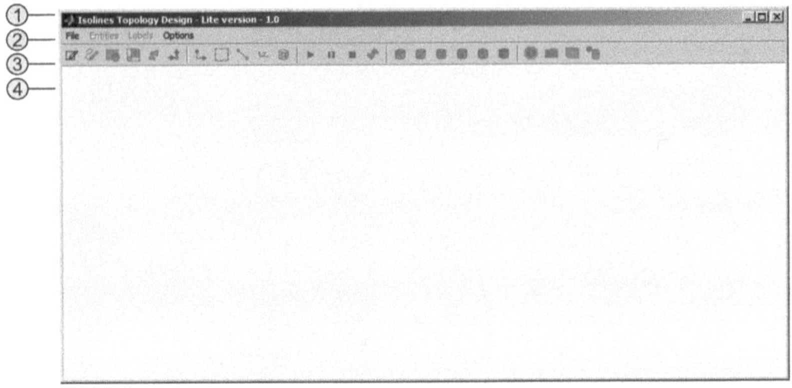

FIGURE 7.33 The liteITD graphical interface.

From top to bottom, the interface consists of four sections: (1) Title bar; (2) Menu bar; (3) Button bar; and (4) Display area. Users can resize the overall size of the display area. To resize this section, either drag the borders around the window or position the mouse on one of the corners of the GUI and drag it diagonally toward the center of the GUI while holding down the left mouse button. Note that, not all menus and buttons are enabled throughout the work session. Each menu and button becomes enabled/disabled depending on which stage of the work session the user is currently in.

7.6.2.1 Title Bar

The Title bar (part 1 in Fig. 7.33) displays the name of the application "Isolines Topology Design – Lite version" together with the version number "1.0". On the right hand side, it also has the standard Windows buttons to minimize, maximize, and close the application.

7.6.2.2 Menu Bar

Each menu topic on the Menu bar (part 2 in Fig. 7.33) brings up a pull-down menu of subtopics, which in turn either cascades to a submenu (denoted by the >) or performs an action. The action may do one of the following: (1) Immediately execute a command; (2) Execute/activate a function (denoted by the ✓); and (3) Bring up a dialog box (denoted by the . . .). Note that, during a session the user will need to use the menu bar buttons from left to right. For this reason, these will switch automatically between enabled and disabled modes depending on which stage of the work session the user is currently in. The liteITD workspace consists of all input data in a liteITD session. *Note that the workspace information is not maintained during different work sessions of liteITD. When the user exits liteITD, the workspace is cleared, although the user can save all input data to the MAT-file (.mat) format for retrieving later.*

The Menu bar has four menus: (1) Files; (2) Entities; (3) Labels; and (4) Options. They are explained in the following sections.

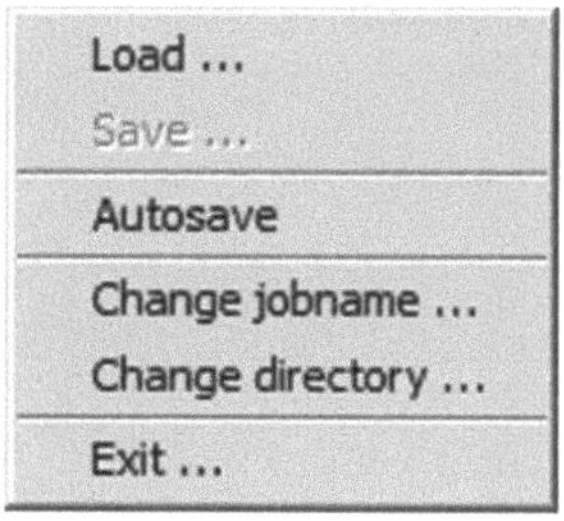

FIGURE 7.34 File drop-down menu list.

When *"File"* is selected, the drop-down menu list of Fig. 7.34 is displayed. There are six commands in the File menu: (1) Load; (2) Save; (3) Autosave; (4) Change jobname; (5) Change directory; and (6) Exit. The actions carried out by each menu command are described next.

1. *Load*: This command restores data from the workspace file. Select the MAT-file in the current folder, left-click, and then select open. The user can load or resume MAT-files at any stage of the work session. A resume operation replaces the data currently in memory with the data in the named workspace file.
2. *Save*: This command saves all input data in a file called "jobname.mat". It is a good practice to save the job at different times throughout the building of the model to backup the work in case of a system failure or other unforeseen problems.
3. *Autosave*: If selected (denoted by the ✓), liteITD automatically saves at each *"step"* the workspace to a file called "jobname_asv-step.mat". The label *"step"* helps the user understand at which stage of the work session the workspace was saved. There are six "*step*" labels associated with the six stages of the work session: (1) *wb* (workbench); (2) *geo* (geometry); (3) *mat* (material properties); (4) *mesh* (mesh of finite elements); (5) *sup* (support conditions); and (6) *for* (forces).

   The *"jobname"* is the name that identifies the liteITD job. When the user defines the jobname, this becomes the first part of the name of all of the files created in that session. By using a different jobname for each work session, the user ensures that no files are overwritten. When the user does not specify a jobname, all files receive the default name *"file"* (Fig. 7.35).
4. *Change jobname*: This command allows the user to specify the jobname for the work session. Type the new jobname in the available field of the dialog box (Fig. 7.35).
5. *Change directory*: This command allows the user to set the folder (by file selection dialog box) where liteITD stores all of the files created during a work session and is also known as the working directory. If the user doesn't specify a working directory, liteITD defaults all outputs to the *outITD* directory.

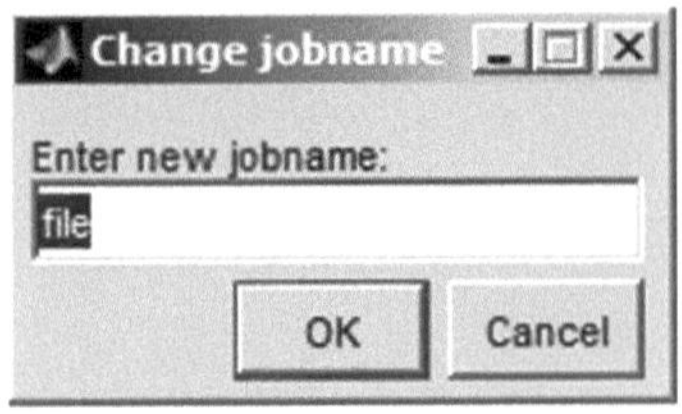

FIGURE 7.35 Change Jobname dialog box.

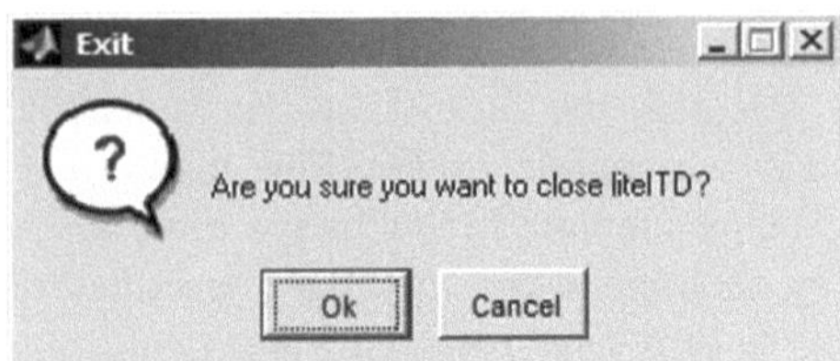

FIGURE 7.36 Dialog box to exit liteITD.

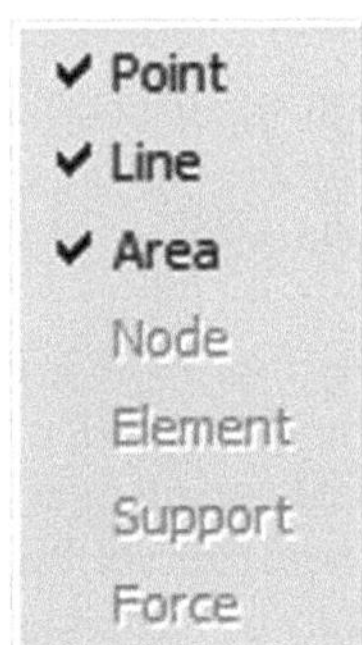

FIGURE 7.37 Entities drop-down menu list.

6. *Exit*: This command will exit the program (Fig. 7.36).

When "*Entities*" is selected, the drop-down menu list of Fig. 7.37 is displayed. There are seven commands in the Entities menu: (1) Point (points); (2) Line (lines); (3) Area (areas); (4) Node (nodes of the finite elements); (5) Element (finite elements); (6) Support (support conditions); and (7) Force (forces). When an entity is enabled, it is selectable in the normal manner. When it is disabled it will be greyed out and this means that there are no entities of that type, as is shown in Fig. 7.37 for Node, Element, Support, and Force. By clicking on the enabled items (checked items have a tick mark ✓ against them), the user can select those entities of the model which are displayed.

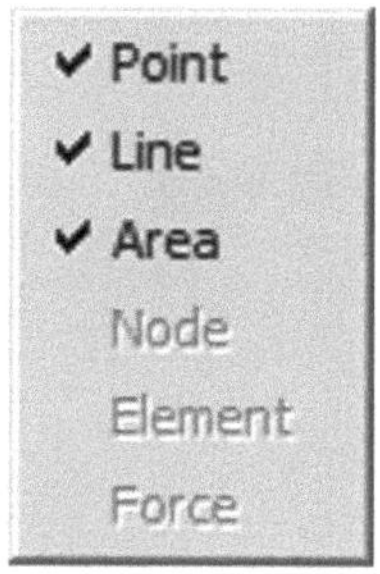

FIGURE 7.38 Labels drop-down menu list.

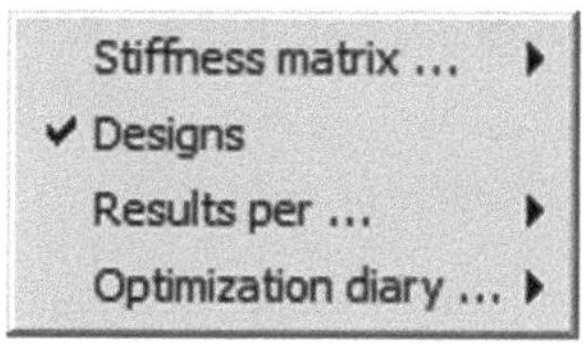

FIGURE 7.39 Options drop-down menu list.

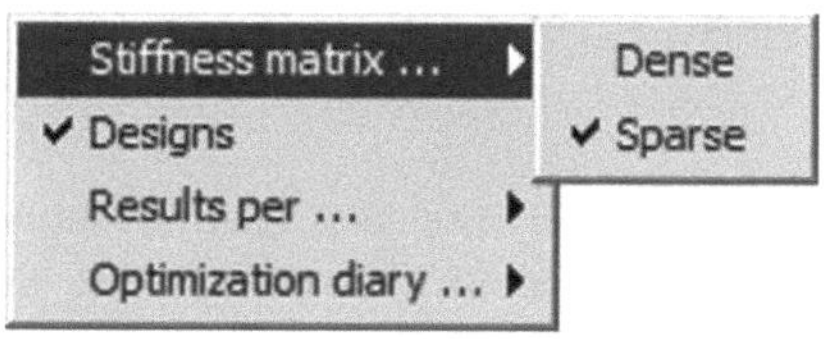

FIGURE 7.40 Stiffness matrix submenu.

When "*Labels*" is selected, the drop-down menu list of Fig. 7.38 is displayed. There are six commands in the Labels menu: (1) Point (point numbers, at the right of items); (2) Line (line numbers, at the middle of items); (3) Area (area numbers, at the center of items); (4) Node (node number, at the right of items); (5) Element (finite element numbers, at the center of items); and (6) Force (force values written at the right of arrows). By clicking on the enabled items, the user can control those entity labels which are displayed.

When "*Options*" is selected, the drop-down menu list of Fig. 7.39 is displayed. There are four commands in the Options menu: (1) Stiffness matrix; (2) Designs; (3) Results per; and (4) Optimization diary. The actions carried out by each menu command are described next.

1. *Stiffness matrix:* When "*Stiffness matrix*" is selected, the submenu of Fig. 7.40 is displayed. There are two options: (1) Dense (full matrix) and (2) Sparse (sparse matrix).

   It is not uncommon to have matrices with a large number of zero-valued elements. Because MATLAB stores zeros in the same way it

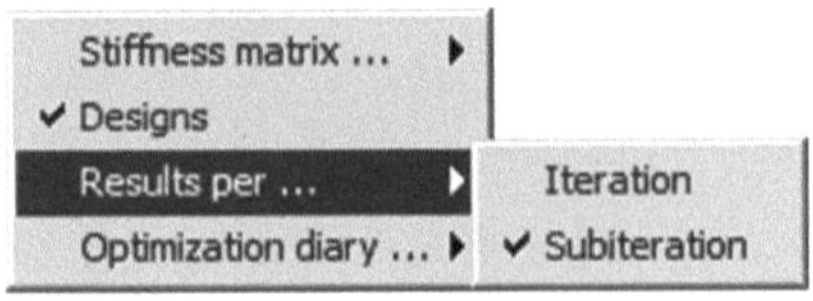

FIGURE 7.41 Result per submenu.

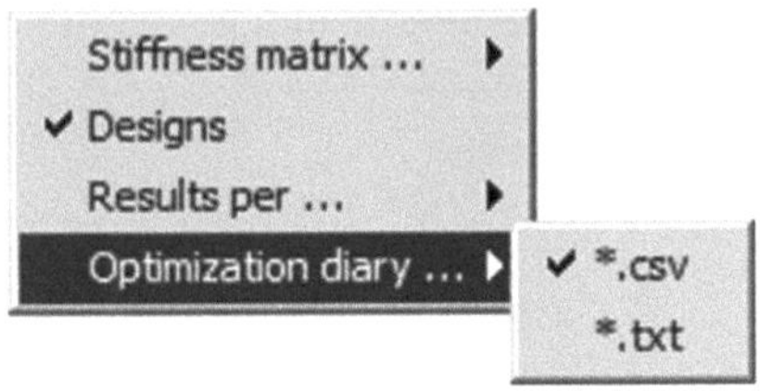

FIGURE 7.42 Optimization diary submenu.

stores any other numeric value, these elements use unnecessary memory space and can require extra computing time. Sparse matrices provide a way to store data with a large percentage of zero elements more efficiently. While full matrices store every element in memory regardless of the value, sparse matrices can significantly reduce the amount of memory required for data storage. For this reason, *Sparce* is the default selected option (Fig. 7.40).

2. *Designs:* When "*Designs*" is checked (default option, Fig. 7.39), the resulting designs shown in the display area during the optimization process are saved to image bitmap files with the extension "*.bmp*". The name of these files is "*jobname_it-%it_sIt-%sIt.bmp*", where "*jobname*" is the jobname, "%it" is the iteration number and "%sIt" is the stabilization or subiteration number.
3. *Result per:* This allows the user to specify the frequency of the resulting designs to display on the screen and to save on the image bitmap files. When "*Result per*" is selected, the submenu of Fig. 7.41 is displayed. There are two frequency options available are: (1) Iteration and (2) Subiteration (*the default option*).
4. *Optimization diary:* This allows the user to export the results of the optimization to the file named "*jobname_optEvol*." When "*Optimization diary*" is selected, the submenu of Fig. 7.42 is displayed. There are two file format options: (1) *.csv (comma separated values text files) and (2) *.txt (delimited text files, in which the TAB character typically separates each field of text). Note that, if the optimization result file is a "*.csv*" file (*the default option*), Microsoft Excel (or similar programs) can automatically open the file and displays the data in a new workbook. Whereas if the file is a "*.txt*" text file, Excel will need to launch the Import Text Wizard.

```
It sIt nAnl timeOpt frVol maxDX maxDY minVM meanVM maxVM
1 0 1 5.243293e-01 0.9817 -5.913135e-07 -1.930125e-06 1.145839e+03 4.074569e+04 1.404917e+05
1 1 2 1.319950e+00 0.9802 -5.852926e-07 -1.736785e-06 8.987488e+03 4.238846e+04 1.404917e+05
2 0 3 1.937809e+00 0.9618 5.851441e-07 -1.736824e-06 8.913492e+03 4.238598e+04 1.404917e+05
2 1 4 2.435620e+00 0.9575 5.811636e-07 -1.743623e-06 8.980456e+03 4.264713e+04 1.404917e+05
3 0 5 2.971723e+00 0.9383 -5.747266e-07 -1.744528e-06 1.394023e+04 4.324693e+04 1.404917e+05
3 1 6 3.601063e+00 0.9308 4.546590e-07 -1.740759e-06 1.197101e+04 4.366425e+04 1.405925e+05
4 0 7 4.111499e+00 0.9087 4.437910e-07 -1.612558e-06 1.649981e+04 4.424509e+04 1.406430e+05
4 1 8 4.714457e+00 0.8900 4.305871e-07 -1.619818e-06 1.606383e+04 4.479970e+04 1.418084e+05
4 2 9 5.238792e+00 0.8742 4.602203e-22 -1.489171e-06 1.716913e+04 4.569078e+04 1.437429e+05
4 3 10 5.873483e+00 0.8588 -3.229694e-07 -1.499056e-06 1.829609e+04 4.649300e+04 1.455201e+05
```

FIGURE 7.43 Format of the results optimization file.

FIGURE 7.44 The Main button menu list.

After the optimization is completed, the optimization result file can be found in the working directory. A small sample of a *".txt"* result file showing the first four iterations of an optimization run is shown in Fig. 7.43. The first and second columns of data identifies the iteration (*It*) and subiteration (*sIt*) numbers, respectively. The third column displays the total number of finite element analysis (*nAnl*). The fourth column shows the total time spent by the optimization in seconds. The fifth column shows the design volume fraction. The last five columns show the values of the maximum displacement in *X*- and *Y*-directions *(maxDX, maxDY)*, and the minimum, mean, and maximum von Mises stresses *(minVM, meanVM, maxVM)*. These values are determined only from those nodes which lie inside or on the boundary of the resulting designs. In cases of multiple loading conditions or different material properties in tension and compression, the last five columns are not saved in the result file.

7.6.2.3 Button Bar

The Button bar (part 3 in Fig. 7.33) has 25 buttons and is divided into five sections: The five sections are named: (1) Main; (2) View; (3) Run; (4) Material; and (5) Miscellaneous. When the user moves the mouse pointer over any button in the Button area, a *tooltip string* will appear to give an explanation of the button functions. This section explains the actions carried out by all of these buttons.

Main Toolbar

The *"Main"* button section consists of the six buttons of Fig. 7.44. These correspond to the following six commands: (1) Define workbench dimensions; (2) Draw geometry; (3) Define material properties; (4) Define mesh; (5) Input support conditions; and (6) Input nodal forces. These buttons are explained below.

1. *Define workbench dimensions button:* Selecting this button  opens the *"Workbench"* dialog box of Fig. 7.45, which allows the user to specify

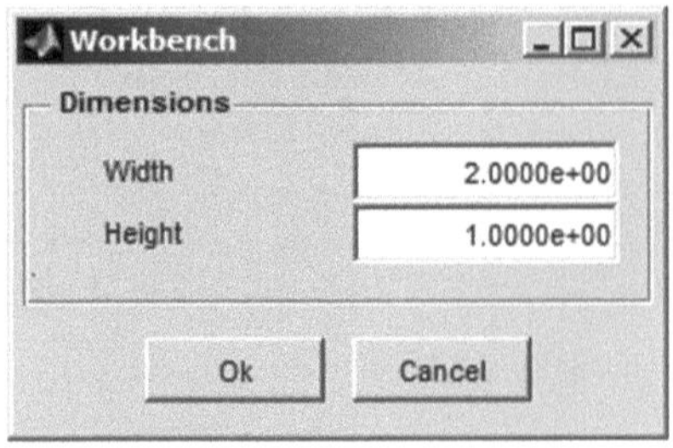

FIGURE 7.45 Workbench dimensions dialog box.

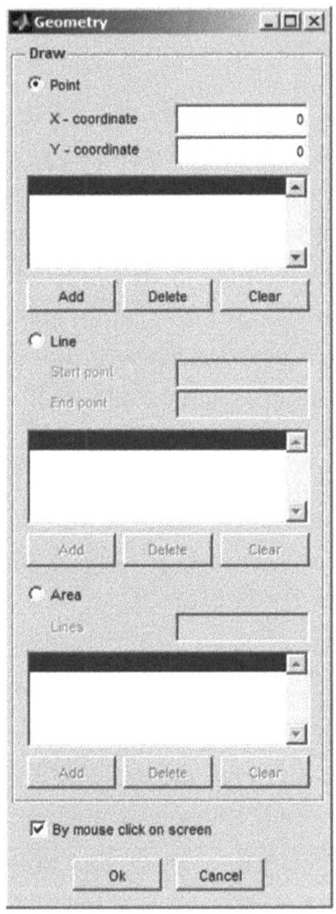

FIGURE 7.46 Geometry dialog box.

the width and height of the workbench. The liteITD program defaults to the SI system of units for the work session. The user can, however, use any system of units, but care must be taken to make all units consistent for all of the input data. See Section 7.9 for an appropriate set of equivalent units. After the workbench dimensions are specified, the user can then draw the geometry of the model.

2. *Draw geometry button:* Selecting this button opens the *"Geometry"* dialog box of Fig. 7.46, which allows the user to *"Draw"* the geometric model of the structure to be optimized.

To obtain an analysis model, it is necessary to generate a finite element model (nodes and finite elements). There are two strategies to create such an analysis model: (1) Direct generation and (2) Solid modeling. With *direct generation*, the user manually has to define the location of each node and the connectivity of each element. With *solid modeling*, the user describes the geometric model, then instructs the program to automatically generate a mesh made from nodes and finite elements within the specified geometry. The solid modeling strategy is used by liteITD. Any plane geometric model is defined in terms of points, lines, and areas. Points are the vertices, lines are the edges, and areas are the faces. There is a hierarchy for these entities: areas, the highest-order entity, are bounded by lines, which are in turn bounded by points.

The user can create ("*Add*" button), remove ("*Delete*" and "*Clear*" buttons), and select points, lines, and areas from the *list boxes* in the "*Geometry*" dialog box of Fig. 7.46. To manage these entities, the user must use the radio buttons "Point," "Line," and "Area." Alternatively the user can add, select, and unselect an entity or location by clicking on it in the display area, provided that the *"By mouse click on screen"* check box has been checked. These four aspects of the "*Geometry*" dialog box are explained below:

a. *Point radio button:* Points are defined by coordinates, and two options can be used: Option (1) by typing the coordinates in the edit boxes (*X*- and *Y*-coordinates, Fig. 7.46) and then selecting the "*Add*" button. Option (2) by directly pressing on the display area (the coordinates of the mouse pointer are automatically shown in the edit boxes) with the left button of the mouse (*Pick Mode* must be enabled, item (d) of this list). This radio button can also be used to select a point: Option (1) by clicking in the point list with the left button of the mouse; Option (2) by picking on the screen with the right mouse button. When a point is selected, this is highlighted in the *point list* and on the screen (using a square marker of blue colour). The "*Delete*" button removes the selected point from the list, and "*Clear*" button removes all points. When a point is removed, the associated bounded lines and areas are also removed. When any entity is deleted, a reordering procedure is made to avoid numeration jumps.

b. *Line radio button:* Lines are defined by two end points and two options can be used: Option (1) by typing the "*Start*" and "*End*" points which define the line in the edit boxes (Fig. 7.46) and then selecting the "*Add*" button; Option (2) by picking in the display area with the left mouse button the two points which defines the line. This radio button can also be used to select a line: Option (1): by clicking in the line list with the left button of the mouse; Option (2) by picking on the screen with the right mouse button (at the middle of the line). When a line is selected, this is highlighted in the line list and also on the screen (using a ticker blue line). The "*Delete*" button removes the selected line from the list, and the "*Clear*" button removes all lines. If a line is removed, bounded areas are also removed.

c. *Area radio button:* Areas are defined by lines and two options can be used: Option (1) by typing "*Line/s*" in the edit box (Fig. 7.46) and then selecting the "*Add*" button; (2) by picking in the display area with the left button of the mouse (at the center of a line) and then selecting the "*Add*" button. This radio button can also be used to select an area: Option (1) by clicking in the area list with the *left* mouse button; Option (2) by picking on the screen with the *right* mouse button. When an area is selected, this area is highlighted in the area list and also on the screen (using a blue background). The "*Delete*" button removes the selected area from the list, and the "*Clear*" button removes all areas. *Two things to note: (1) The lines which define an area do not need to be specified in consecutive order (although, in the area list they are sorted counter clockwise). (2) When there is a large number of lines to input, MATLAB provides a special shortcut notation by using the colon operator. This operator specifies a series of values by specifying the first value (first) in the series, the stepping increment (incr) and the last value (last) in the series. For example, the expression 1:2:7 (first:incr:last) is a shortcut for a 1 × 4 row vector containing the values 1,3,5 and 7.*

d. *By mouse click on screen Check Box:* This check box ☑ By mouse click on screen is used to activate the "*Pick Mode*," which allows the user to add, select, and unselect an entity or location by using the mouse to click on the display area. The default setting is checked. A summary of the mouse-button assignments used during a picking operation is given below:

  i. Left mouse button adds the entity or location closest to the cursor of the mouse pointer.

  ii. Right mouse button selects the entity or location closest to the click location of the mouse pointer.

  iii. Double-click any mouse button unselects the selected entities.

The mechanical properties of the materials used for the structure can only be specified after drawing the geometric model.

3. *Define material properties button:* Selecting this button  opens the "*Material properties*" dialog menu of Fig. 7.47, which allows the user to specify the mechanical properties of the materials for the analysis model. The user can specify the elastic (Young's) modulus and Poisson's ratio for each material. The user can define up to five different material types. The total number of materials is specified in the "*Number of materials*" pop-up menu. Each set of material properties has a material reference number "*No.*" and can be allocated one of five colours: cyan, black,

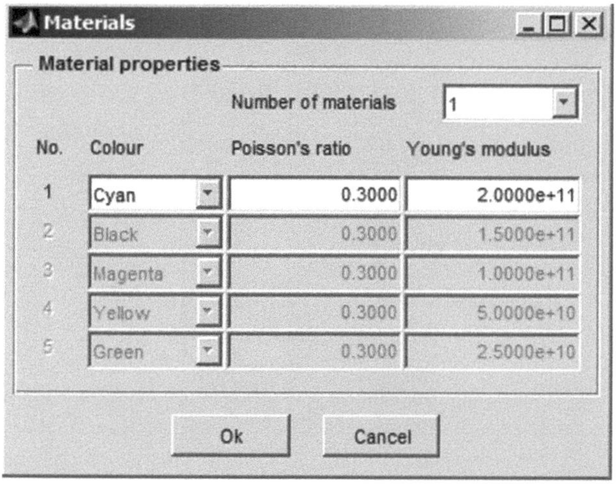

FIGURE 7.47 Materials Property Dialog box for specifying the material properties.

magenta, yellow, and green. *Please note that: (1) The unit system must be consistent, the default material properties are SI units (Pa); (2) The Poisson's ratios must be in the range of 0–0.5; and (3) The elastic modulus must be positive.*

The process of meshing the geometric model with a finite element mesh, can only be carried out after the material properties have been defined.

4. *Define mesh button:* Selecting this button opens the "*Mesh*" dialogue menu of Fig. 7.48. The menu is divided into six sections: (a) Fixed grid; (b) Shape; (c) Type; (d) Size; (e) Attributes; and (f) By mouse click on screen Check Box. These are explained below.
   a. *Fixed grid:* This edit box allows the user to specify the Fixed Grid ratio (FG_R) between the outside (O) and inside (I) material properties of the fixed grid mesh, as explained in Section 4.5.1 and defined by Eq. (7.12). Typical values of this ratio are in the range of $10^{-6} < FG_R < 10^{-4}$. The default value for *liteITD* is 1×10^{-4}.

$$\mathbf{K}_O = FG_R \times \mathbf{K}_I \tag{7.12}$$

   b. *Shape*: This section allows the user to specify the type of finite element shape to be used for the mesh generation. Two different FE shapes are available by selecting from the radio buttons: (1) "Tri" which are triangular shaped finite elements and (2) "Quad" which are quadrilateral shaped finite elements. *Note that a mixture of the two element types is not supported by this program.*
   c. *Type*: This section allows the user to specify the type of meshing that will be used by the program to mesh the geometric model. Two types

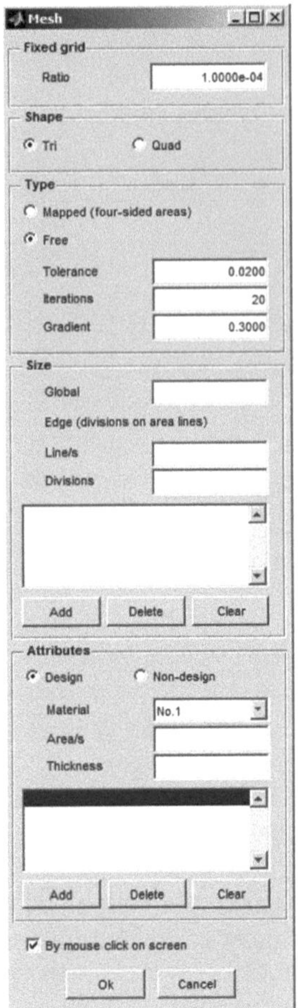

FIGURE 7.48 Mesh Dialog box.

are available by selecting from the radio buttons: (i) "*Mapped*" and (ii) "*Free*." The user should consider before meshing the model, if a *mapped* or *free* mesh is appropriate for the FEA and for the optimization. These two methods are described below:

i. *Mapped*: This type of mesh builds the geometry as a series of fairly regular areas that can accept a mapped mesh. For an area to accept a mapped mesh, two conditions must be satisfied: (1) the area must be bounded by four lines (four-sided areas) and (2) the area must have equal number of element divisions specified on opposite sides. *Note that: (a) The program can only create a mapped mesh within convex 4-sided regions; (b) The program generates a mapped triangle mesh starting from a mapped quadrangle mesh, dividing each quadrangle into two triangles.*

   ii. *Free:* This type of mesh allows the modeling of any type of geometry of the solid model, even if this is irregular. When the "*Free*" option is selected, the edit boxes named: "*Tolerance*", "*Iterations*", and "*Gradient*" are activated. These options allow the user to modify the default behaviour of the free mesh generator, where "*Tolerance*" is the converge tolerance (the maximum relative change in edge length per iteration must be less than this value, which defaults to 0.02). "*Iterations*" is the maximum allowable number of iterations, which defaults to 20, and "*Gradient*" is the maximum allowable (relative) gradient in the size function, which defaults to 0.3. Note that, the program generates a free quadrangle mesh starting from a free triangle mesh by quadrangulation (by joining the triangle centroids to the midpoints of its sides, three quadrilaterals are obtained for each of these triangles).

d. *Size:* This section controls the element size used to create the mesh. The user can choose from either the "*Global*" option to specify the maximum allowable global element size (maximum edge length), or the "*Edge*" option to specify the number of divisions on lines (shortcut notation is allowed), or both options together. In this latter case, the program will use the option that provides the smallest element size. The list box in this panel group, allows the user to select lines or manage (add, delete, or clear) the size specifications.

e. *Attributes*: This section allows the user to specify the area attributes, which must be specified before the mesh is generated. These attributes consist of:

   i. "*Design*" or "*Non-design*": These radio buttons allow the user to specify the type of domain. "*Design*" domain corresponds to all areas which are allowed to take part in the optimization. "*Non-design*" domain are areas which are required as part of the structure, but which cannot be optimized.

   ii. "*Material*": This drop-down menu allows the user to select the material property number to be attributed to an area.

   iii. "*Area/s*": This edit box allows the user to type the number of the areas associated with the above mentioned area attributes. Alternatively the areas to which the attributes are given can be selected by picking on the screen with the *right* mouse button.

   iv. "*Thickness*": This edit box allows the user to specify the thickness of the selected material property associated with the selected "*Area/s*."

   v. The list box in this panel group, allows the user to select the areas or manage the area attributes (add, delete, or clear).

f. *By mouse click on screen Check Box:* This check box 

is used to activate the "*Pick Mode*," which

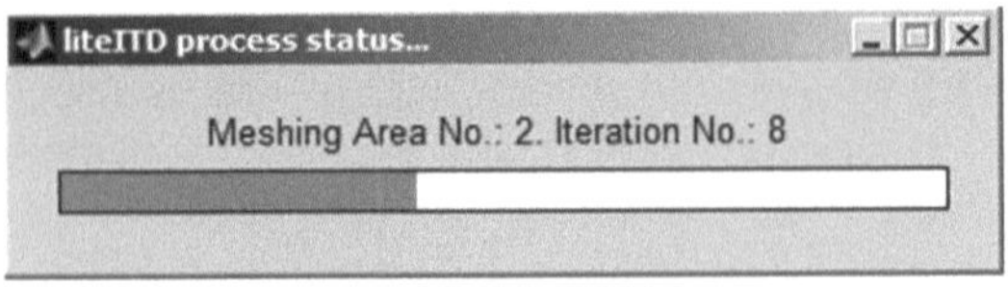

FIGURE 7.49 Meshing process status dialog box. The text shows the number of the area being meshed and the number of completed iterations.

allows the user to add, select, and unselect an entity by using the mouse to click on the display area. The default setting is checked. A summary of the mouse-button assignments used during a picking operation is given below:

- **i.** Left button adds to the "*Line/s*" edit box the line closest to the mouse cursor.
- **ii.** Right button adds to the "*Area/s*" edit box the area closest to the mouse cursor. *Note that the mouse cursor must be over the label of the area for selection, i.e., A1, A2, etc.*
- **iii.** Double-click any mouse button unselects any selected entities.

Selecting the "*OK*" button in the "*Mesh*" dialogue menu of Fig. 7.48, will close this menu and start to generate the mesh. If the mesh type "*Free*" was selected, the "*liteITD process status*" dialog box of Fig. 7.49 appears. It shows the current percentage completed status of the meshing process, by progressively filling the rectangular bar with red (black in print versions) from left to right.

The process of applying the support conditions, can only be carried out after the geometric model has been meshed.

5. *Input support conditions button:* Selecting this button opens the "*Displacements on nodes*" dialog box of Fig. 7.50. It allows the user to specify constrained displacements on the nodes. The dialog box is divided in three sections: (a) Constrained freedom; (b) Selection; and (c) By mouse click on screen Check Box. These are explained below.

a. *Constrained freedom:* The user can only apply Degrees of Freedoms (DOF) constraints on nodes. Three different DOF constraint conditions can be selected:

- **i.** Displacement in the *X*-direction is fixed (*X*-displacement box is checked);
- **ii.** Displacement in the *Y*-direction is fixed (*Y*-displacement box is checked);
- **iii.** All displacements are fixed (both the *X*- and *Y*-displacement boxes are checked).

b. *Selection:* The user can select a node by typing its number in the "*Node/s*" edit box (MATLAB shortcut notation can be used) or by selecting it graphical if the "*Pick Mode*," has been activated, part (c)

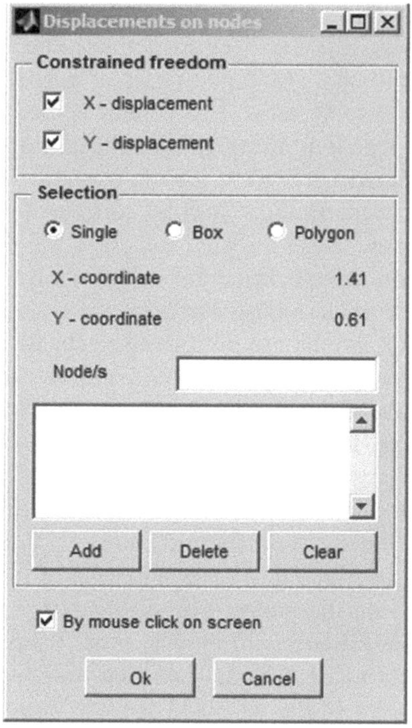

FIGURE 7.50 Displacement on Nodes dialog box.

of this list. When using the "*Pick Mode*," there are three options: "*Single*", "Box", and "*Polygon*" mode:

i. *Single pick mode:* In this mode, each click on the left mouse button selects one node.
ii. *Box pick mode:* In this mode, the user moves the mouse to a point on the work bench to define the start of the box. Click once the left mouse button, then drag the mouse to enclose a set of nodes in a box and click on the left mouse button again to select the box of nodes.
iii. *Polygon pick mode*: In this mode, the user moves the mouse to a point on the work bench to define the start of the polygon. Click once the left mouse button, then drag the mouse to the next polygon point and click on the left mouse button again. This process is repeated until the polygon is drawn to select the nodes inside of it.

When a node or a group of nodes are selected, they are highlighted using a square marker of blue colour. When the "*Add*" button is pressed, the DOF constraints specified in (a) above are applied to the selected nodes and are shown in the list box. When the "*Delete*" button is pressed, the constrained nodes are removed from the list. When the "*Clear*" button is pressed, all constrained nodes are removed from the list.

c. *By mouse click on screen Check Box:* This check box ☑ By mouse click on screen is used to activate the "*Pick Mode*," which allows the user to add, select and unselect a node by using the mouse to click on the display area. The default setting is checked. A summary of the mouse-button assignments used during a picking operation is given below:

   i. Left button adds to the "*Node/s*" edit box the node closest to the mouse cursor.

   ii. Right button selects from the nodes shown in the list box, the node closest to the mouse cursor.

   iii. Double-click any mouse button unselects all selected nodes.

   Note that, when *Pick Mode* is active, the mouse cursor coordinates are continually updated in the "*X-coordinate*" and "*Y-coordinate*" label section as the mouse is moved over the workbench.

   The process of applying the loading conditions, can only be carried out after the support conditions have been specified.

6. *Input nodal forces button:* Selecting this button opens the "*Forces on nodes*" dialog box of Fig. 7.51. It allows the user to apply the forces (or concentrated loads) on the nodes. The dialog box is divided in four sections: (a) Load case number; (b) Nodal force; (c) Selection; and (d) By mouse click on screen Check Box. These are explained below.

   a. *Load case number:* In the edit box, the user needs to specify the total number of load cases. After which, the "*Apply*" button needs to be pressed. This will then activate the, "*Nodal force*" and "*Selection*" panels.

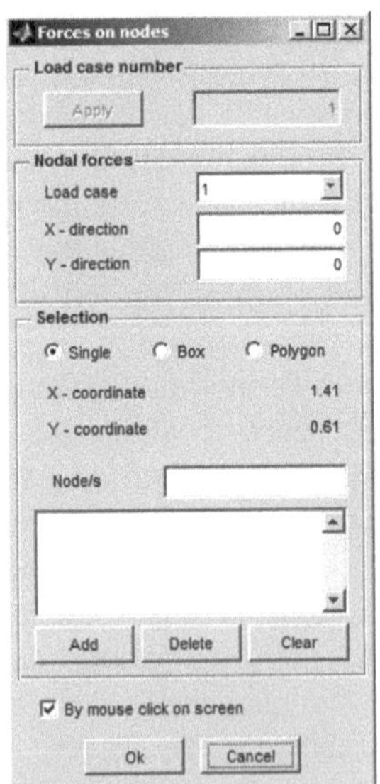

FIGURE 7.51 Dialog box for specifying loading conditions.

b. *Nodal forces:* the user can choose which load case is current by left-clicking on the "*Load case*" pop-up menu. The magnitude of the applied force is specified in the "*X-direction*" and "*Y-direction*" edit boxes. The user needs to apply forces which are of a consistent unit system (Section 7.9).

c. *Selection:* The user can select a node by typing its number in the "*Node/s*" edit box (MATLAB shortcut notation can be used) or by selecting it graphical if the "*Pick Mode*," has been activated, part (d) of this list. When using the "*Pick Mode*," there are three options: "*Single*", "Box," and "*Polygon*" mode:

   i. *Single pick mode:* In this mode, each click on the left mouse button selects one node.

   ii. *Box pick mode:* In this mode, the user moves the mouse to a point on the work bench to define the start of the box. Click once the left mouse button, then drag the mouse to enclose a set of nodes in a box and click on the left mouse button again to select the box of nodes.

   iii. *Polygon pick mode*: In this mode, the user moves the mouse to a point on the work bench to define the start of the polygon. Click once the left mouse button, then drag the mouse to the next polygon point and click on the left mouse button again. This process is repeated until the polygon is drawn to select the nodes inside of it.

   When a node or a group of nodes are selected, they are highlighted using a square marker of blue colour. When the "*Add*" button is pressed, the nodal forces specified in (b) above are applied to the selected nodes and are shown in the list box. When the "*Delete*" button is pressed, the forces are removed from the list. When the "*Clear*" button is pressed, all forces are removed from the list.

d. *By mouse click on screen Check Box:* This check box 

is used to activate the "*Pick Mode*," which allows the user to add, select and unselect a node by using the mouse to click on the display area. The default setting is checked. A summary of the mouse-button assignments used during a picking operation is given below:

   i. Left button adds to the "*Node/s*" edit box the node closest to the mouse cursor.

   ii. Right button selects from the nodes shown in the list box, the node closest to the mouse cursor.

   iii. Double-click any mouse button unselects all selected nodes.

Note that, when *Pick Mode* is active, the mouse cursor coordinates are continually updated in the "*X-coordinate*" and "*Y-coordinate*" label section as the mouse is moved over the workbench.

View Toolbar

The "*View*" button section consists of the five buttons of Fig. 7.52. These correspond to the following five commands: (1) Show triad; (2) Show workbench; (3) Show entities; (4) Show labels; and (5) Display edges of finite elements. These buttons as explained below.

1. *Show triad button:* When this button is in the pressed down mode, it shows the global XY coordinate triad in the lower left hand corner of the display area.
2. *Show workbench button:* When this button is in the pressed down mode, it shows the workbench dimensions using a dashed black rectangle.
3. *Show entities button:* When this button is in the pressed down mode, it shows the entities marked with the tick symbol (✓) in the "*Entities*" menu.
4. *Show labels button:* When this button is in pressed down mode, it shows the labels marked with the tick symbol (✓) in the "*Labels*" menu.
5. *Display edges of finite elements button:* When this button is in pressed down mode, it displays the boundaries (edges) of the finite elements with a solid black line.

Optimization Run Toolbar

The "*Optimization Run*" button section consists of the four buttons of Fig. 7.53. These correspond to the following four commands: (1) Optimization launcher; (2) Optimization pause; (3) Optimization stop; and (4) Restart liteITD program. These buttons as explained below.

1. *Optimization launcher button:* Selecting this button launches the "*Optimization*" dialog box of Fig. 7.54. It allows the user to specify the *liteITD* parameters. The values shown in this figure are the default settings specified by *liteITD* for all designs. The dialog box is divided in four sections: (a) Material; (b) Load case; (c) T-ension/C-ompression; and (d) Evolution.
   a. *Material*: In this section the user can choose from a single or a multi-material design. When the "*Single*" material option is selected, the

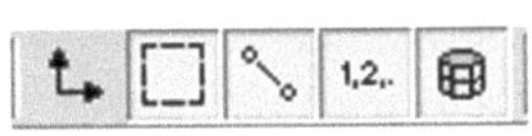

FIGURE 7.52 The View button menu list.

FIGURE 7.53 The Optimization Run button menu list.

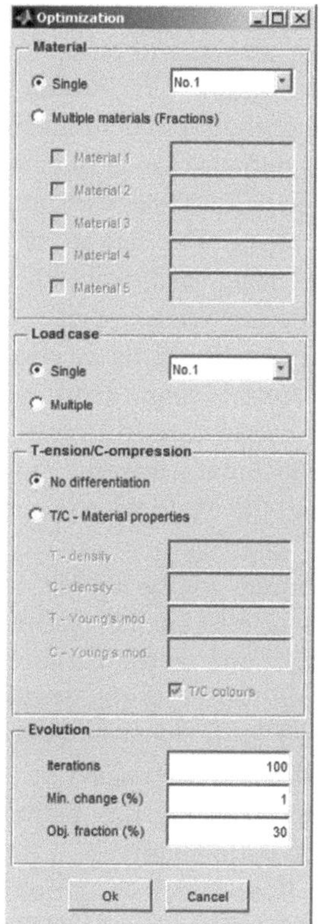

FIGURE 7.54 Optimization dialog box.

user must specify a material number from the pop-up menu. When the "*Multiple materials*" option is chosen, the user needs to use the Material 1–5 check boxes to select the materials and the volume control weighting factors used to optimize the structure, Section 4.7.3. *Note that when the "Single" material option is selected, the "Multiple materials (Fractions)" options are disabled.*

b. *Load case:* In this section the user can choose from single or multiple load cases conditions. When the "*Single*" option is selected, the user must select the load case to optimize the structure with respect to, from the pop-up menu. When the "*Multiple*" option is selected, the optimization is carried out for multiple load cases, Section 4.7.2.

c. *T-ension/C-ompression:* This section allows the user to select the necessary parameters to optimize a topology with different material properties in tension and compression, Section 4.3.2. To do this, the "*T/C-material properties*" radio button must be selected. This then requires the user to specify the following information: (a) The density of tension and compression members (T-density, C-density) and (b) the elastic modulus of the tensile and compressive members (T-Young's mod., C-Young's mod.). The "*T/C colours*" check box, allows the user to select to display the regions in tension and compression in red and blue colours. The default case is this box checked.

d. *Evolution*: This section allows the user to set the evolutionary parameters. These consist of: (a) The number of iterations ("Iterations"); (b) The minimum volume change as a percentage ("Min. change (%)"); and (c) the objective volume fraction as a percentage ("Obj. fraction (%)").

Note that the topology optimization is allowed only in the following four cases: (1) Single material, single load case, no differentiation T/C; (2) Single material, single load case, different material properties in tension and compression; (3) Single material, multiple loading conditions, no differentiation T/C; and (4) Multimaterials, single load case, no differentiation T/C.

Selecting the "*Ok*" button in the dialog box starts the optimization process. The computational cost of the optimization will depend on the size of finite element model and on the number of load cases.

2. *Optimization Pause button:* Selecting this button causes the optimization process to be paused until the user presses the button again, thus allowing the optimization to be continued.

3. *Optimization Stop button:* Selecting this button causes the optimization process to be stopped permanently.

4. *Restart liteITD Program button:* Selecting this button causes all input data to be permanently deleted and the liteITD program is restarted.

Material Toolbar

The "*Materials*" button section consists of the six buttons of Fig. 7.55. The first five buttons (from left to right) allow the user to filter what is displayed in the design domain by material number. The last button allows the user to show/hide the regions treated as nondesign domain in the resulting designs. Note that: (1) The number of enabled buttons to filter by material number needs to be consistent with the number of materials used in the optimization and (2) The "*Display nondesign regions*" is enabled irrespective of there being nondesign regions present in the model.

Miscellaneous Buttons

The "*Miscellaneous*" button section consists of the four buttons of Fig. 7.56. These correspond to the following four commands: (1) Show information;

FIGURE 7.55 The Materials button menu list.

(2) Screenshot; (3) Raise hidden menu; and (4) Active dynamic model manipulation. These buttons as explained below.

1. *Show information button:* When this button  is in the pressed down mode, relevant information is shows on the display area for the user (Fig. 7.57). This is split into two areas:
   a. *On the left side of the screen:* Is shown the following information (from top to bottom, Fig. 7.57A): Working directory (*Working directory*); jobname (*Jobname*) ; time; date; number of points (*Points*), lines (*Lines*), areas (*Areas*), and materials (*Materials*); fixed grid ratio (*FG ratio*); number of nodes (*Nodes*), elements (*Elements*), degree of freedoms (*DoFs*), constrained degree of freedoms (*Const. DoFs*); number of load cases (*Load cases*); number of materials (*Opt. materials*) and load cases (*Opt. load cases*) that take part in the optimization; tension and compression factor (*T/C factor*); number of iterations (*Iterations*); minimum volume change (*Min. change*); and objective volume fraction (*Obj. Fraction*).

FIGURE 7.56 The Miscellaneous Buttons menu list.

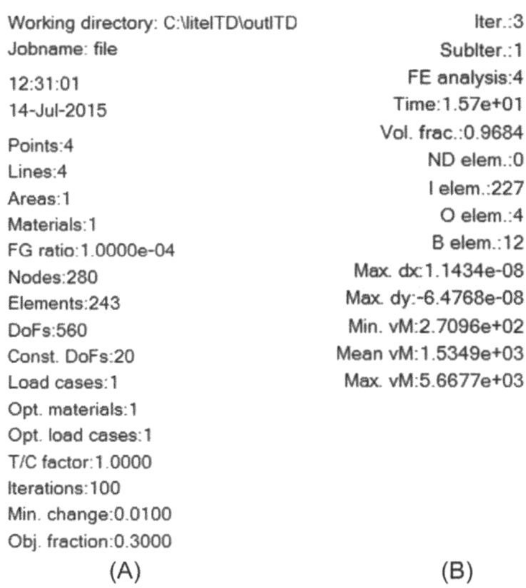

FIGURE 7.57 Information shown on the display area during the optimization: (A) Information about the model, (b) Information about the optimization solution.

   b. *On the right side of the screen:* Is shown the following information (from top to bottom, Fig. 7.57B): Current iteration (*Iter*) and subiteration (*SubIter*) number; number of finite element analysis (*FE analysis*); elapsed time (*Time*); current volume fraction (*Vol. frac.*); number of nondesign elements (*ND elem*), inside (*I elem*), outside (*O elem*), and boundary (*B elem*) elements; maximum displacement in X- (*Max. dx*) and Y- (*Max. dy*) directions; minimum (*Min. vM*), mean (*Mean vM*), and maximum (*Max. vM*) von Mises stress of the nodes which lie inside or on the boundary of the design domain.

   Note that depending on the stage of the optimization cycle, some of the information of Fig. 7.57 may not be displayed.

2. *Screenshot button:* Selecting this button allows the user to save the display area to an image bitmap file with the extension "*.bmp*". The file is named "*jobname_screenshot_%iShot*," where "*jobname*" is the jobname and "*%iShot*" is the capture number.
3. *Raise hidden menu button*: Selecting this button allows the user to raise a hidden window or dialog box to the top of liteITD graphical interface. *Note that sometimes a menu window or dialog box may inadvertently not be able to be seen because it is hidden. Selecting this button will allow the user to see that window. It is advisable not to work with liteITD in the full Window mode. Instead, it is best to have the graphical interface of* Fig. 7.33 *occupy about ¾ of the width of the computer screen. Then all dialog boxes which are selected can occupy the remaining ¼ of the width of the computer screen. Then the case of a "hidden" window or dialog box will never occur.*
4. *Active dynamic model manipulation button*: Selecting this button activates the dynamic model manipulation mode. The user can use the mouse pointer to pan or zoom the model.
   a. Press and hold down the left mouse button to pan the model.
   b. Using the mouse scroll wheel lets user zoom in/out.
   c. Double-click with any mouse button resets the view to the initial default view size.

Note that when the "*By mouse click on screen*" check box is checked (see *Geometry, mesh, displacements* and *forces on nodes pop-up menus*) the dynamic model manipulation mode is disabled.

7.7 STEP-BY-STEP GUIDE TO USE liteITD

This section provides a step-by-step guide on how to use liteITD to generate and then optimize a topology optimization model. The procedure includes the following eight steps:

1. Define the design workbench dimensions.
2. Draw the geometric model.

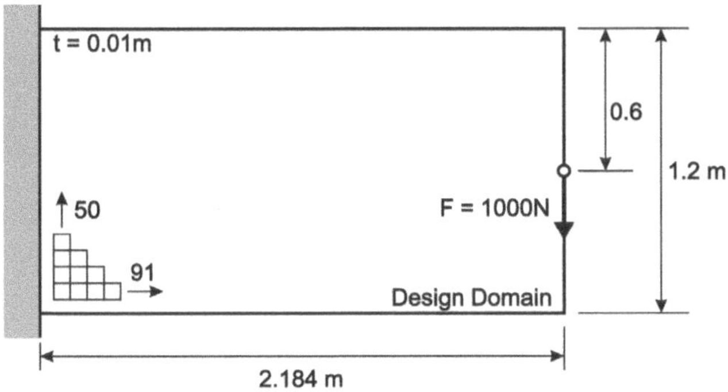

FIGURE 7.58 Design domain for Michell cantilever aspect ratio of $\overline{L} = 1.82$.

3. Specify the material properties.
4. Generate the finite element mesh.
5. Apply the DOF constraints.
6. Apply the loading conditions.
7. Specify the liteITD parameters.
8. View the resulting optimal design.

To show how liteITD works, the Michell cantilever [2] of Section 5.2 with an aspect ratio of $\overline{L} = 2.184/1.2 = 1.82$ (Fig. 7.58) was optimized. The design domain consisted of a rectangular area of size 2.184 m × 1.2 m and a thickness of 0.01 m. The mesh used had 91 × 50 elements. A vertically downwards force of magnitude 1000 N was applied at the center of the free end and the cantilever was fully clamped along the left edge. The elastic modulus of all elements was set to 200×10^9 Pa, the Poisson's ratio was 0.3, and the fixed grid ratio Eq. (7.12) was set to the default value of the program $FG_R = 10^{-4}$. The ITD parameters were: total number of iterations: 50; final volume fraction: 50%; and minimum volume change: 1%.

1. Create a directory inside of the *outITD* directory on the Desktop called *Michell_Cantilever* (Fig. 7.59).
2. Start liteITD.exe.
3. Click on "*Autosave*" (*Menu bar* > *File* > *Autosave*).
4. Open "*Change jobname*" (*Menu bar* > *File* > *Change jobname...*). When the dialog box of Fig. 7.35 appears, type "*Michell_Cantilever_SLC*" and press "*OK*" (Fig. 7.60).
5. Open "*Change directory*" (*Menu bar* > *File* > *Change directory...*) When the file selection dialog box appears, select the directory *Michell_Cantilever*. All working files will now be saved in this directory.

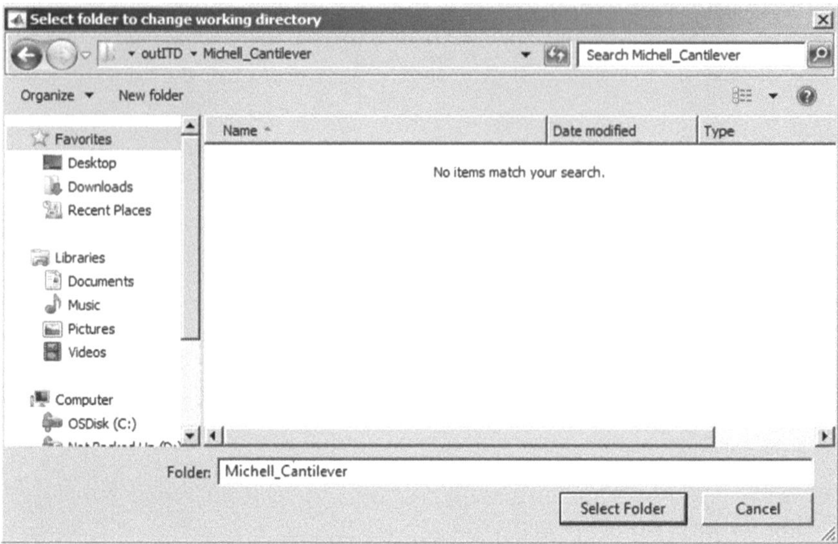

FIGURE 7.59 Create the Michell_Cantilever directory to set as the working directory.

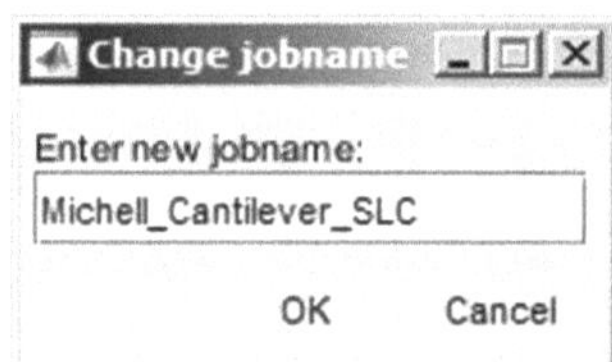

FIGURE 7.60 Change jobname dialog box showing the new jobname.

7.7.1 Define the Design Workbench Dimensions

To define the design workbench dimensions involves the following two steps:

1. Press the "*Define workbench dimensions*" button from the main toolbar.
2. When the "*Workbench*" dialog box of Fig. 7.45 appears, enter "*2.184*" in the *"Width"* edit box and enter *"1.2"* in the *"Height"* edit box, then press "*Ok*" (Fig. 7.61).
3. Press the "*Show information*" button from the main toolbar. The resulting workbench of Fig. 7.62 is then drawn in the display area.

7.7.2 Draw the Geometric Model

After the workbench is defined, the following five steps show how to generate the geometric model. In this case, a rectangular-shaped area.

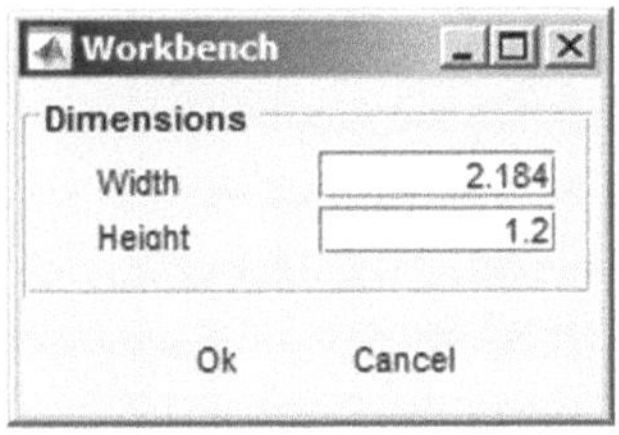

FIGURE 7.61 Workbench dimensions dialog box showing the 2.184 x 1.2 m workbench.

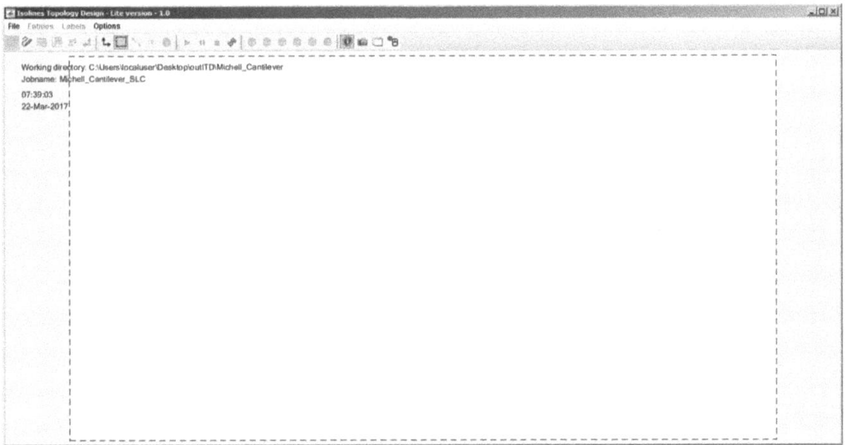

FIGURE 7.62 liteITD graphical interface showing drawn specified workbench.

1. Press the "Draw geometry" button in the main toolbar, causing the "Geometry" dialog box of Fig. 7.46 to appear. Note that, when the user creates/deletes entities such as points, lines, or areas, these are automatically numbered.
2. Pick the location of the corners of the workbench or type the coordinates of the points as follows: enter "0" in the "*X*-coordinate" and "0" in the "*Y*-coordinate" edit boxes, then press the "Add" button. Repeat this procedure with the remaining three points of coordinates (*x*, *y*): (2.184, 0); (2.184, 1.2); (0, 1.2).
3. Select the "Line" radio button. Pick the points "1" and "2" or type 1 in the "Start point" and 2 in the "End point" edit boxes, then press the "Add" button. Repeat the procedure to generate the next 3 lines by selecting their end points: "2" and "3"; "3" and "4"; "4" and "1."
4. Select the "*Area*" radio button. Pick the lines "L1", "L2", "L3", and "L4" or type, e.g., "1:4" in the "Line/s" edit box, then press the "Add" button (Fig. 7.63). The grey area is numbered as "A1" (Fig. 7.63).
5. Press "Ok." The expected geometry is shown in Fig. 7.64.

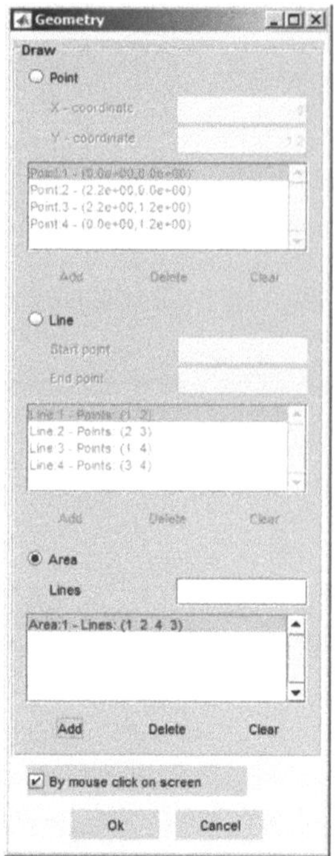

FIGURE 7.63 Geometry dialog box showing the geometry data.

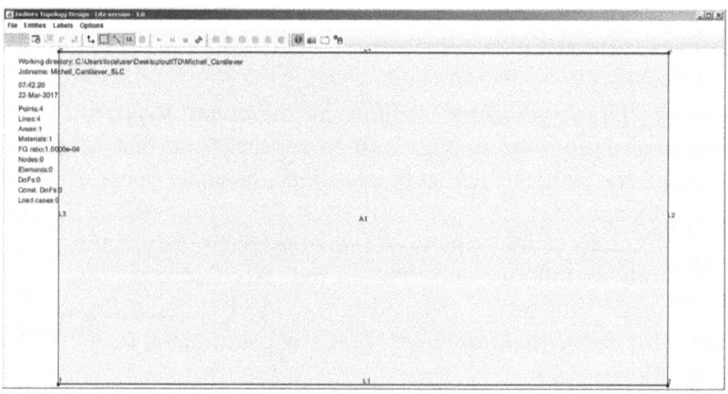

FIGURE 7.64 liteITD graphical interface showing the drawn geometric model.

7.7.3 Specify the Material Properties

Press the "*Define material properties*" button located in the main toolbar and the "*Materials*" dialog box of Fig. 7.47 will appear with the same default values. For this example, the default values need not be changed. Just press the "*Ok*" button (Fig. 7.65).

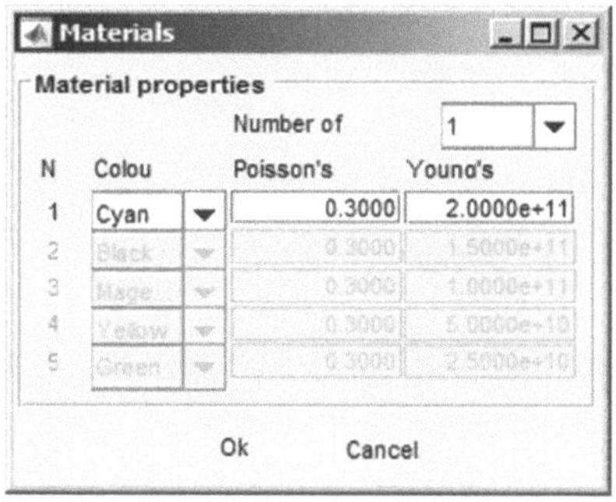

FIGURE 7.65 Material property defined for the Michell cantilever.

7.7.4 Generate the Finite Element Mesh

After the material properties are defined, the geometric model is ready to be meshed. From within the main toolbar, press the "*Mesh generator*" button. The "*Mesh*" dialog box of Fig. 7.48 will appear. Follow the following five steps:

1. From the "*Shape*" panel, select the "*Quad*" option.
2. From the "*Type*" panel, select the "*Mapped (four-sided areas)*" option.
3. From the "*Size*" panel:
   a. Pick or type in "*Line/s*" edit box "1 4" and in "*Divisions*" edit box "91," then press "*Add*" button;
   b. Pick or type in "*Line/s*" edit box "3 2" and in "*Divisions*" edit box "50," then press "*Add*" button.
4. From the "*Attributes*" panel, pick the area by clicking with the right mouse button *directly over* the name A1 on the workspace or type in "*Area/s*" edit box "1" and in the "*Thickness*" edit box type "*0.01*," then press the "*Add*" button. The user can review or manage the input data using the list boxes included in the dialog box (Fig. 7.66).
5. Press the "*Ok*" button. When the process is completed, the mesh of Fig. 7.67 will be drawn.

7.7.5 Apply the DOF Constraints

After the geometric model is meshed, the user can now define the DOF constraints of the nodes. By clicking on the "*Input support conditions*" button, the "*Displacements on nodes*" dialog box will appear (Fig. 7.50). Follow the following four steps:

1. From "*Selection*" panel, select the "*Box*" radio button.
2. Move the mouse into the display area to the upper left screen corner. At this location, click the left mouse button, and move to the right and down

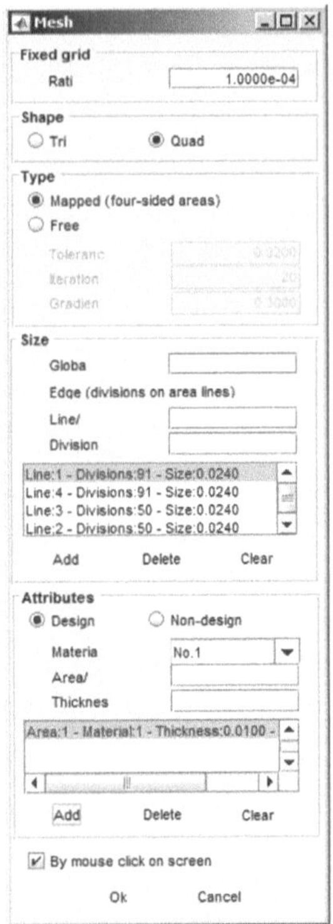

FIGURE 7.66 Mesh dialog box showing the mesh data.

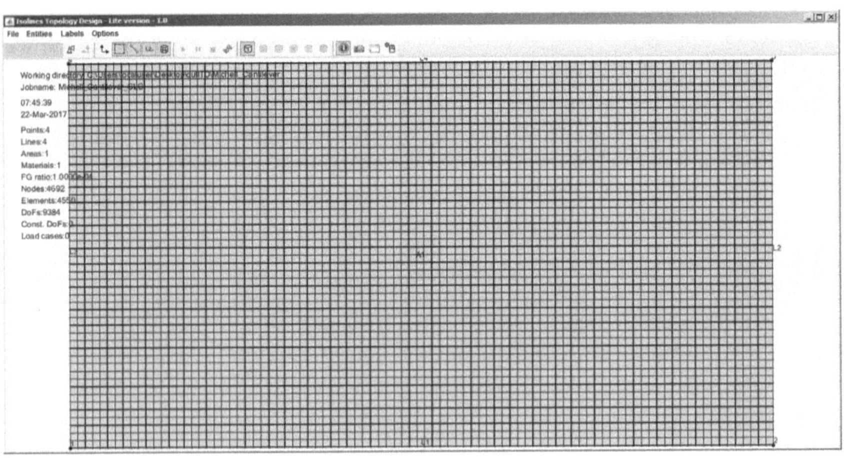

FIGURE 7.67 liteITD graphical interface showing the generated finite element mesh.

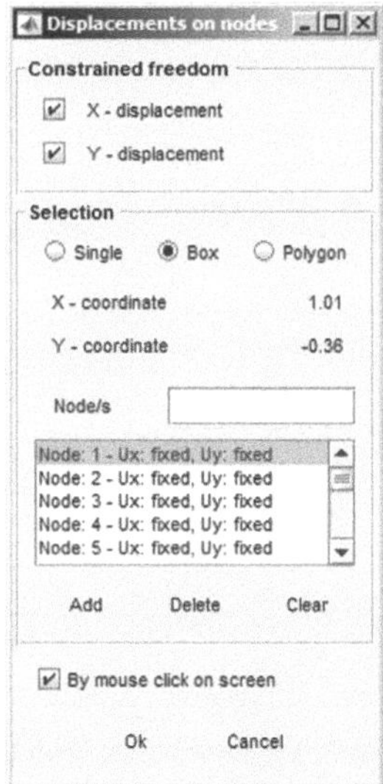

FIGURE 7.68 Displacements on nodes dialog box showing the DOF constraint data.

until the first column of nodes is inside of the dashed blue box, then click the left mouse button again. The selected nodes are marked using a blue square marker and their labels are shown in the "*Node/s*" edit box.

3. Press the "*Add*" button. The selected nodes will be included in the list box (Fig. 7.68), and a two triangles (*X- and Y-displacements are fixed*) per constrained node will show in the display area (Fig. 7.70).
4. When done, press the "*Ok*" button.

7.7.6 Apply the Loading Conditions

To specify forces at the nodes, the user needs to select the "*Input nodal forces*" button, which will launch the "*Forces on nodes*" dialog box of Fig. 7.51. Follow the following four steps:

1. From the "*Load case number*" panel, press "*Apply*" button. For this example, the number of load cases need not be changed.
2. Input the force components as "*0*" for "*X-direction*" and " − *1000*" for "*Y-direction*."
3. Move the mouse to the position *(2.184, 0.6)*, the mouse position is displayed on the "*Selection*" panel. Press the left button to include the node nearest to the mouse location in the "*Node/s*" edit box.
4. Press "*Add*" button. The selected node will be included in the list box (Fig. 7.69), and a red arrow (black arrow in print versions) pointing down will be drawn on the mesh (Fig. 7.70).

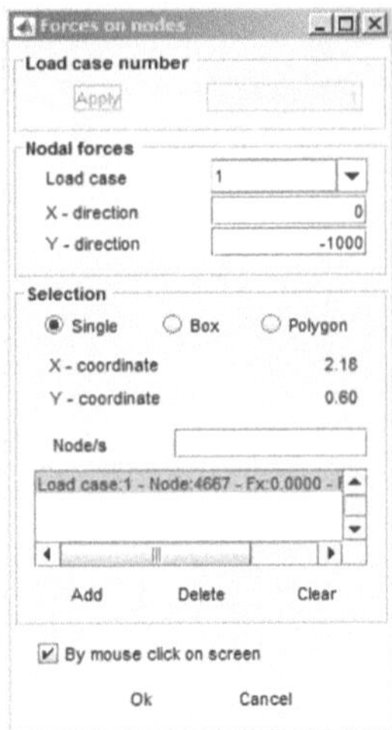

FIGURE 7.69 Forces on nodes dialog box showing the loading conditions data.

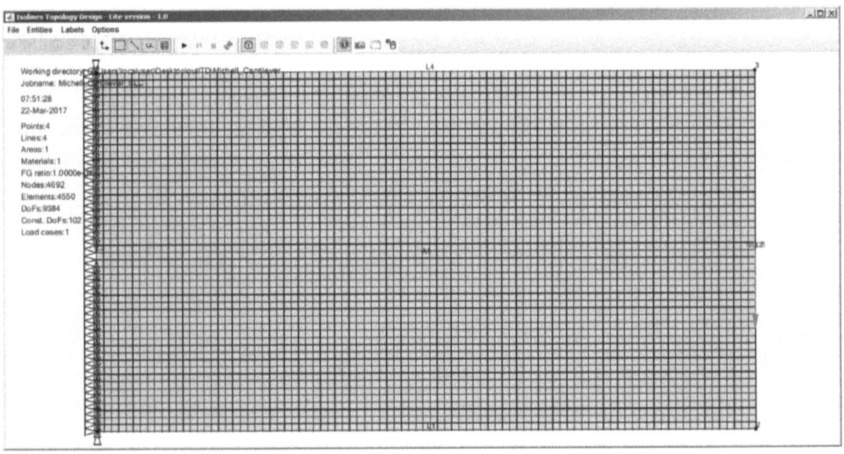

FIGURE 7.70 liteITD graphical interface showing the support and loading conditions.

7.7.7 Specify the liteITD Parameters and Run Optimization

Press the "*Optimization launcher*" button. The dialog box of Fig. 7.54 will appear. From the "*Evolution*" panel, in the "*Iterations*" edit box replace "*100*" with "*50*," in the "*Obj. fraction (%)*" edit box replace the default value of "*30*" with "*50*." Press the "*Ok*" button to accept these new parameters and to start the optimization.

FIGURE 7.71 liteITD workbench showing the final optimal design for the Michell cantilever spect ratio of $\overline{L} = 1.82$ and 50% volume fraction after 50 iterations and 74 structural analysis.

7.7.8 View the Resulting Optimal Design

After fifty iterations, the optimization run is finished and the final design with a 50% volume fraction will be obtained as shown in Fig. 7.71.

7.8 ADDITIONAL LITEITD EXAMPLES

In order to illustrate the other optimization algorithms of Chapter 4, Continuous Method of Structural Optimization, implemented in the liteITD program, the Michell cantilever of aspect ratio of $\overline{L} = 1.82$ was optimized a further three times for the following other conditions:

1. Multiple loading conditions;
2. With different material properties in tension and compression; and
3. Using multiple-materials.

7.8.1 Michell Cantilever under Multiple Loading Conditions

Most real structures are subjected to different loads at different times (not all loads act simultaneously). This is referred to as multiple loading conditions (e.g., a moving load can suitably be simulated by multiple load cases). With multiple loading conditions, the resulting design has to be optimized to account for all load cases.

The geometric model, material properties, mesh characteristics, and support conditions used in this example are the same as for the first example (Section 7.7). The upward force $F_1 = 1000$ N is applied at the upper right-hand corner and represents load case 1. The downward force $F_2 = 1000$ N is applied at the lower right-hand corner and represents load case 2 (Fig. 7.72). The ITD parameters used were: total number of iterations: 50; final volume fraction: 40%; minimum volume change: 1%.

To carry out the topology optimization for multiple loading conditions for the Michell cantilever, the follow 12 steps need to be followed:

1. Start "liteITD.exe".
2. Open "Change jobname" dialog box, then type "Michel_Cantilever_MLC" and press "OK."

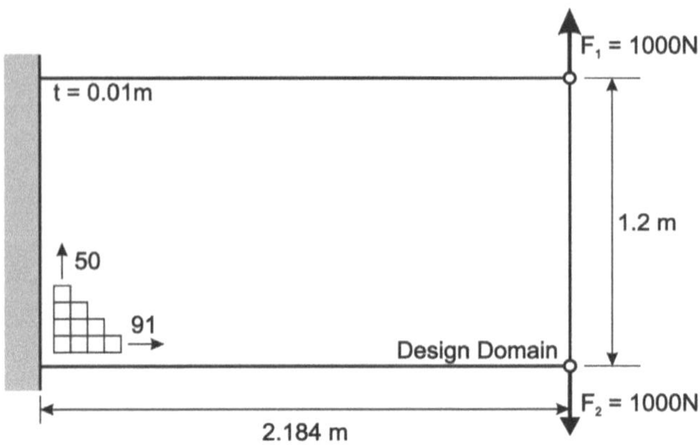

FIGURE 7.72 Design domain for the Michell cantilever of aspect ratio of $\overline{L} = 1.82$ under two load cases, where F_1 and F_2 represent each load case.

3. Repeat Sections 7.7.1−7.7.5 or load the "Michel_Cantilever_SLC_asv-sup.mat" file.
4. Press the "Input nodal forces" button. Type "2" in the "Load case number" edit box, then press "Apply" button.
5. From the "Nodal forces" panel, type "1000" in the "*Y*-direction" edit box.
6. Select the node located at the upper right-hand corner, then press "Add" button.
7. Select "2" from the "Load case" pop-up menu.
8. From the "Nodal forces" panel, type " − 1000" in the "*Y*-direction" edit box.
9. Select the node located at the lower right-hand corner, then press "Add" button.
10. Press "Ok" button in order for these values to be applied.
11. Press the "Optimization launcher" button. From the "Load case" panel, select "Multiple" radio button.
12. From the "Evolution" panel, in the "Iterations" edit box replace "100" with "50" and in the "Obj. fraction (%)" edit box replace "30" with "40." Press the "Ok" button to accept these new parameters and to start the optimization. After 50 iterations, the final optimal design for the two load case problems is shown in Fig. 7.73.

7.8.2 Michell Cantilever with Different Properties in Tension and Compression

For this model, the material properties, support condition, thickness, and load are exactly the same as for the first example of Section 7.7. The

FIGURE 7.73 liteITD workbench showing the final optimal design for two load cases with a 40% volume fraction, 50 iterations and 158 structural analysis.

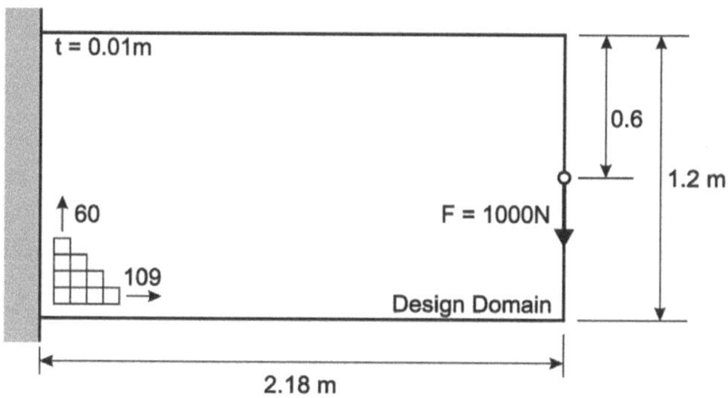

FIGURE 7.74 Design domain for Michell cantilever aspect ratio of $\overline{L} = 1.8167$.

geometric model and mesh characteristics are however slightly different (see Fig. 7.74). The domain aspect ratio was set to $\overline{L} = 2.18/1.2 = 1.8167$, and the mesh used had 109×60 elements. The properties for the materials in tension and compression used were: Elastic moduli $E_T = 200 \times 10^9$ Pa and $E_C = 150 \times 10^9$ Pa, the Poisson's ratio and material density was the same for both: $\nu_T = \nu_C = 0.3$ and $\rho_T = \rho_C = 8000$ kg/m^3. The ITD parameters used were: total number of iterations: 60; final volume fraction: 40%; minimum volume change: 1%.

To carry out the topology optimization for a problem with different tensile and compressive structural behaviour for the Michell cantilever, follow the following nine steps:

1. Start "liteITD.exe".
2. Open "Change jobname" dialog box, then type "Michell_Cantilever_TC" and press "OK."
3. Repeat the step from Section 7.7.1, but use the dimensions of 2.18 for the width and 1.2 for the height.

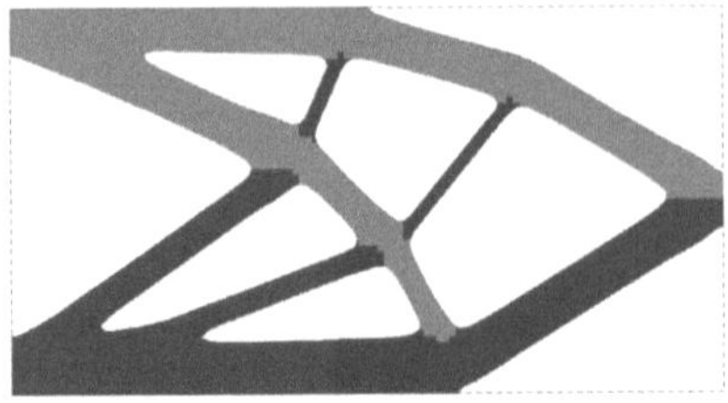

FIGURE 7.75 liteITD workbench showing the final optimal design for the Michell Cantilever with different material properties in tension and compression, a 40% volume fraction, 60 iterations and 101 structural analysis.

4. Repeat the step from Section 7.7.2, but specify the coordinates of the geometry as the following four points (x, y): (0, 0), (2.18, 0); (2.18, 1.2); (0, 1.2).
5. Repeat the step from Section 7.7.3.
6. Repeat the step from Section 7.7.4, but use 109 divisions to subdivide lines 1 and 4, and 60 divisions to subdivide lines 2 and 3.
7. Repeat the step from Section 7.7.5.
8. Repeat the step from Section 7.7.6 but move the mouse to the position (2.18, 0.6).
9. Press the "Optimization launcher" button. From the "T-ension/C-ompression" panel, select "T/C–Material properties" radio button. Then for both "T-density" and "C-density" type "8000." For "T-Young's mod." type "2E + 11" and for "C-Young's mod." type "1.5E + 11." From the "Evolution" panel, in the "Iterations" edit box replace "100" with "60" and in the "Obj. fraction (%)" edit box replace "30" with "40." Press the "Ok" button to accept these new parameters and to start the optimization. The final optimal design is shown in Fig. 7.75. The regions in tension are displayed in red (black in print versions) and those in compression are displayed blue (gray in print versions).

7.8.3 Michell Cantilever Using Multimaterials

The geometric model, mesh characteristics, support conditions, and forces applied for the Michell cantilever are the same as for the example of Section 7.8.2 (Fig. 7.74). In this example, two materials were used, with the elastic modulus of $E_1 = 200 \times 10^9$ Pa and $E_1 = 20 \times 10^9$ Pa, with equal Poisson's ratios of $\nu_1 = \nu_2 = 0.3$ and volume control weighting factors equal to $w_1 = w_2 = 0.5$. The material with the highest stiffness (material No. 1) is displayed in magenta, and the material No. 2 is displayed in cyan. The ITD parameters used were: total number of iterations: 50; final volume fraction: 20%; minimum volume change: 2%.

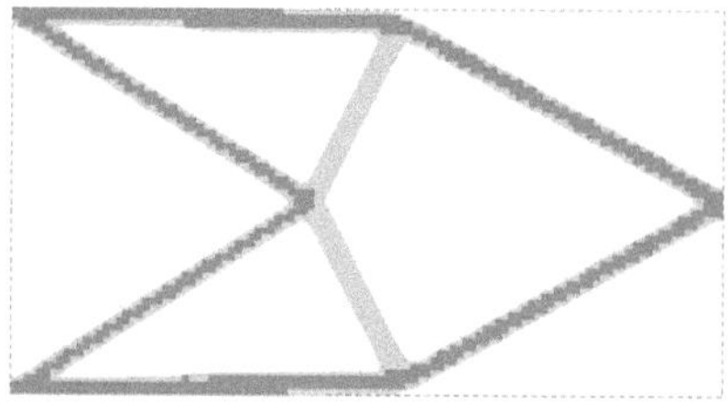

FIGURE 7.76 liteITD workbench showing the final optimal design for the Michell cantilever with two-material schemes and a 20% volume fraction, 50 iterations and 102 structural analysis.

To carry out the topology optimization for a problem using two-material schemes for the Michell cantilever, follow the following eight steps:

1. Start "liteITD.exe."
2. Open "Change jobname" dialog box, then type "Michell_Cantilever_MM" and press "OK."
3. Repeat steps (3) and (4) from Section 7.8.2 or load "Michell_Cantilever_TC_asv-geo.mat" file.
4. Press the "Define material properties" button. From the "Number of materials" pop-up menu, select "2." For the material No. 1, select "Magenta" from the "Colour" pop-up menu. For the material No. 2, select "Cyan" colour and in the "Young's modulus" edit box type "2E + 10." Then Press "Ok."
5. Repeat step (6) of Section 7.8.2 by selecting material 1 for area A1. As the optimization will determine the appropriate material distribution in the structure, at the start of the optimization the entire domain can be meshed using only one material.
6. Repeat steps (7) and (8) of Section 7.8.2.
7. Press the "Optimization launcher" button. From the "Material" panel, select "Multiple materials (Fractions)" radio button. Click on the "Material 1" and "Material 2" check boxes, and then type "0.5" in both edit boxes.
8. From the "Evolution" panel, in the "Iterations" edit box replace "100" with "50," in the "Min. change (%)" edit box replace the "1" with "2" and in the "Obj. fraction (%)" edit box replace "30" with "20." Press the "Ok" button to accept these new parameters and to start the optimization. The final optimal design for two-material schemes is shown in Fig. 7.76.

7.9 APPROPRIATE EQUIVALENT UNITS

The program liteITD requires a consistent system of units to be used for a model. By default liteITD uses SI units. Table 7.5 provides some equivalent units which can be used.

TABLE 7.5 Consistent Sets of Units to be Used in liteITD

Quantity	Units Used			
Length (*L*)	mm	m	in	ft
Force (*F*)	N	N	lbf	lbf
Elastic modulus (*E*)	N/mm^2 (MPa)	N/m^2 (Pa)	lbf/in^2	lbf/ft^2
Density (ρ)	$tonne/mm^3$	kg/m^3	lb/in^3	lb/ft^3
	Conversion Factors Between Units			
L	1000	1	39.36996	3.28083
F	1	1	0.2248089	0.224809
E	10^{-6}	1	1.450377×10^{-4}	0.02088543
ρ	10^{-12}	1	3.612729×10^{-5}	0.06242796

REFERENCES

[1] MathWorks, Inc., 2016. Natick, MA, MATLAB R2016a®.

[2] G.I.N. Rozvany, Exact analytical solutions for some popular benchmark problems in topology optimization, Struct. Optim. 15 (1998) 42–48.

[3] O. Sigmund, A 99 line topology optimization code written in Matlab, Struct Multidisc. Optim. 21 (2001) 120–127.

Index

Note: Page numbers followed by "*f*" and "*t*" refer to figures and tables, respectively.

D

E

F

G

H

I

J

K

L

M

N

O

P

R

S

T

MIX
Papier | Fördert gute Waldnutzung
FSC® C083411

Kolibri 360 GmbH
Ferdinand-Jühlke-Straße 7
99095 Erfurt, Deutschland
produktsicherheit@kolibri360.de